LEARNING GUIDE
FOR
TORTORA
PRINCIPLES
OF
HUMAN ANATOMY

LEARNING GUIDE
FOR
TORTORA'S
PRINCIPLES
OF
HUMAN ANATOMY

THIRD EDITION

Kathleen Schmidt Prezbindowski

College of Mount St. Joseph
and
University of Cincinnati

Gerard J. Tortora

Bergen Community College

1817

HARPER & ROW, PUBLISHERS, New York
Cambridge, Philadelphia, San Francisco,
London, Mexico City, São Paulo, Sydney

Sponsoring Editor: Claudia M. Wilson
Project Editor: H. Detgen
Designer: T. R. Funderburk
Production Manager: Kewal K. Sharma
Compositor: American–Stratford Graphic Services, Inc.
Printer and Binder: The Murray Printing Company
Art Studio: J & R Services, Inc.

Learning Guide for Tortora's *Principles of Human Anatomy,* Third Edition
Copyright © 1983 by Harper & Row, Publishers, Inc.

Library of Congress Cataloging in Publication Data

Prezbindowski, Kathleen S.
   Learning guide for Tortora's Principles of human
anatomy.

   1. Human physiology—Problems, exercises, etc.
2. Anatomy, Human—Problems, exercises, etc. I. Tortora,
Gerard J.   II. Tortora, Gerard J.   Principles of human
anatomy.   III. Title.   [DNLM: 1. Anatomy—Problems.
2. Physiology—Problems.   QS 4 T712pa Suppl.]
QP40.P739   1983   612   82-21164
ISBN 0-06-045291-9

To my parents and sister,
Norine, Harry, and Maureen Schmidt

Contents

To the Student   *ix*

To the Student

This *Learning Guide* is designed to help you do exactly what its title indicates. It will serve as a step-by-step aid to help you bridge the gap between goals (objectives) and accomplishment (learning). The twenty-five chapters of the *Guide* parallel the twenty-five chapters of Tortora's *Principles of Human Anatomy,* Third Edition. Each chapter consists of the following parts:

Introduction. Before you begin your study of each chapter, read this brief overview. It will present the scope of the chapter and will correlate each main topic with related groups of objectives. It will also help to integrate new material with what you have learned in previous chapters.

Topics Summary. This section provides a concise outline of chapter content.

Specific Objectives. The objectives listed at the beginning of each textbook chapter are included here for convenient reference. Objectives organize the learning of the total chapter into manageable tasks. Try to accomplish each individual objective. In this way, you will find that the chapter as a whole will not overwhelm you. (Your instructor may wish to modify the list of objectives according to the goals and time frame of your course.)

Learning Activities. As you move to each new chapter topic, thoroughly read the corresponding text pages (noted in the *Guide*). Then carry out the related activities in the *Guide*. These thought-provoking exercises are designed to increase your depth of understanding of the text and to challenge and verify your knowledge of key concepts. In short, they will assist you, not only in studying, but in *truly learning*.

A variety of learning activities are included. There are subjective excercises such as definitions, comparisons, and paragraphs in which you fill in blanks with key terms. You will also draw, complete, or color figures. (Colored felt-tipped pens will be useful.) Objective activities include matching exercises, multiple-choice questions, and identification of parts of figures.

When you have difficulty with an exercise, refer to the related pages (listed in the margin of the *Guide*) of *Principles of Human Anatomy,* Third Edition before proceeding further.

Answers to Numbered Learning Activities. As you glance through the activities, you will notice that a number of questions are marked with boxed numbers (such as [14]). Answers to such questions are given at the end of each chapter, preceding the Mastery Test, in order to provide you with some immediate feedback about your progress. Incorrect answers alert you to review related objectives in the *Guide* and corresponding text pages.

Activities without boxed numbers and answers are also purposely included in this book. The intention is to encourage you to verify some answers independently by consulting your textbook and also to stimulate discussion with students or instructors.

Mastery Test. This fifteen-question self-test provides an opportunity for a final review of the chapter as a whole. Its format will assist you in preparing for standardized tests or course exams that are objective in nature (multiple choice, true-false, and arrangement questions). Answers for all twenty-five mastery tests are placed together at the end of the book.

I wish you success and enjoyment in *learning* concepts and relevant applications of human anatomy.

K. S. P.

LEARNING GUIDE
FOR
TORTORA'S
PRINCIPLES
OF
HUMAN ANATOMY

1

An Introduction to the Human Body

As you begin your study of *human anatomy*, you will first define this science and recognize its subdivisions (Objective 1). You will then learn about structure-function relationships at different levels of organization within the body (2–4). In order to avoid getting lost during your travels through the human body, you will familiarize yourself with directional terms and planes which relate to the body (6–9) and identify major organs by location (5, 10–11). You will also learn about techniques used to study internal body structure, particularly those used for diagnosis of disease (12–13). You will also study common metric units used in measuring the human body (14).

Topics Summary

A. Anatomy and physiology defined
B. Levels of structural organization
C. Structural plan
D. Radiographical anatomy
E. Measuring the human body

Objectives

1. Define anatomy, with its subdivisions, and physiology.
2. Compare the levels of structural organization that make up the human body.
3. Define a cell, a tissue, an organ, a system, and an organism.
4. Identify the principal systems of the human body, list the representative organs of each system, and describe the function of each system.
5. Describe the basic structural plan of the human body.
6. Define the anatomical position.
7. Compare common and anatomical terms used to describe the external features of the human body.
8. Define directional terms used in association with the human body.
9. Define the common anatomical planes of the human body.
10. List by name and location the principal body cavities and their major organs.
11. Describe how the abdominopelvic cavity is divided into nine regions and four quadrants.
12. Define the relationship of radiographic anatomy to the diagnosis of disease.
13. Point out the differences between a roentgenogram and a CT (computed tomography) scan.
14. Define the common metric units of length, mass, and volume, and their U.S. equivalents, that are used in measuring the human body.

Learning Activities

A. Anatomy and physiology defined

   1. Contrast each of the following pairs of terms:
     a. Anatomy/physiology

     b. Gross anatomy/histology

     c. Systematic anatomy/regional anatomy

     d. Developmental anatomy/embryology

   2. Look at the two items in each of the following pairs. Tell how their structural differences explain their functional differences.
     a. Spoon/fork

     b. Hand/foot

     c. Chair/bed

     d. Incisors (front teeth)/molars

B. Levels of structural organization

   1   1. Rearrange these terms in order from lowest to highest level of organization, using lines provided.

Term	Level of organization
Organism	(lowest)
Cell	
Tissue	
Organ	Organ
Chemical	
System	(highest)

2. Complete the following table describing systems of the body. Name two or more organs in each system. Then list one or more functions of each system.

System	Organs	Functions
a.	Skin, hair, nails	
b. Skeletal		
c. Muscular		
d.	Blood, heart, blood vessels	
e. Lymphatic		
f.		Regulates body by nerve impulses
g.	Glands that produce hormones	
h.		Supplies oxygen, removes carbon dioxide, regulates acid-base balance
i.		Breaks down food and eliminates solid wastes
j.	Kidney, ureters, urinary bladder, urethra	
k. Reproductive		

C. Structural plan

(pages 10–18)

1. Write three anatomical characteristics that describe the structural plan of the human body.

4

2 2. Using your own body, a skeleton, a torso, and/or Figure 1-3 in your text, determine relationships among body parts. Write the correct directional term(s) to complete each of these statements.

    a. The liver is _____ to the diaphragm.

    b. Fingers (phalanges) are located _____ to wrist bones (carpals).

    c. Another word for *dorsal* is _____ .

    d. The great (big) toe is _____ to the little toe.

    e. The little toe is _____ to the great toe.

    f. The skin on your leg is _____ to muscle tissue in your leg.

    g. Muscles of your arm are _____ to skin on your arm.

    h. When you lie face down in a pool, as if to do the "deadman's float," you are lying on your _____ surface.

    i. The lungs and heart are located _____ to the abdominal organs.

3. Complete the table summarizing body cavities.

Body Cavity	Divisions	Organs Contained
a. Dorsal		
b.	Thoracic	

4. Which of these cavities is better protected? (*Dorsal? Ventral?*) Of what advantage is this?

5. The thorax is further subdivided into three regions. Two of these are called the _____ cavities; they contain the lungs. Located between them is the region named the _____ . One important structure located here is the _____ . Check Figure 1-7 in your text to verify your answers.

3 6. The abdominopelvic cavity has two subdivisions. Answer these questions about them.

    a. Which region is more superior in location? (*Abdominal? Pelvic?*)

    b. Which region contains the urinary bladder, rectum, and internal reproductive organs? (*Abdominal? Pelvic?*)

    c. What imaginary line demarcates the two cavities? (See Figure 1-6 in your text.)

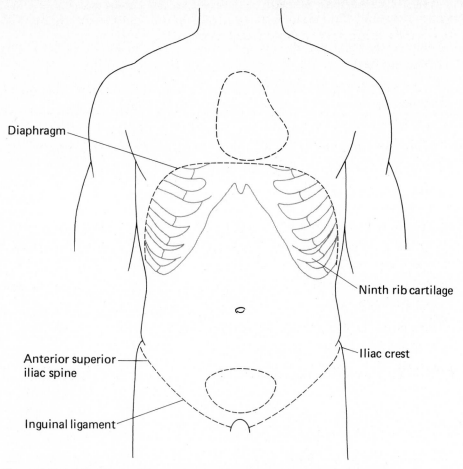

Diaphragm

Ninth rib cartilage

Iliac crest

Anterior superior
iliac spine

Inguinal ligament

Figure LG 1-1 Regions of the ventral body cavity. Complete as directed.

7. Complete Figure LG 1-1 according to directions below.
   a. Draw and label thoracic organs.
   b. Draw lines dividing the abdomen into nine regions. After carefully studying Figure 1-8 in your text, try to draw and label from memory the following organs in their correct locations: *stomach, liver, colon of large intestine, appendix,* and *small intestine.* Then draw and label the *kidneys* in contrasting color (since these organs are posterior to other abdominal viscera).
8. Using the names of the nine abdominal regions, complete these statements. Refer to the figure you just completed and to Figure 1-8 in your text.
   a. From superior to inferior, the three abdominal regions on the right side

      are: right _____ , right _____ ,

      and right _____ .
   b. The stomach is located primarily in the two regions named

      _____ and _____ .

   c. The navel is located in the _____ region.
   d. The region immediately superior to the urinary bladder is named the

      _____ region.
   e. If the abdomen were divided into four quadrants, the left lower quadrant

      would include all of the _____ region, and parts of

the three regions named _____ , _____

_____ , and _____ .

9. Complete the table relating common terms to anatomical terms. (See Figure 1-2 in your text to check your answers.)

Common Term	Anatomical Term
a.	Axillary
b. Fingers	
c. Arm	
d.	Popliteal
e.	Cephalic
f. Mouth	
g.	Inguinal
h. Chest	
i.	Cervical

10. Describe the *anatomical position.*

11. Match each of the following planes with the phrases telling how the body would be divided by such a plane.

[4]

_____ a. Into superior and inferior portions

_____ b. Into equal right and left portions

_____ c. Into anterior and posterior portions

_____ d. Into unequal right and left portions

F. Frontal
H. Horizontal
M. Midsagittal
S. Sagittal

(pages 18–21)   **D.** Radiographic anatomy

1. Briefly describe the method used to produce a roentgenogram by means of x-rays.

2. Explain what advantage is offered by a CT scan rather than a roentgenogram for diagnostic purposes.

2

Cells: Structural Units of the Body

In this chapter you will examine structure and function at the cellular level. You will look first at the general organizational plan of these living building blocks (Objective 1). Then you will study the manner in which materials enter or leave cells (2–3). You will look closely at the cell's interior and see the role of each organelle (4–15). You will learn how cells divide (16–17). Finally, you will contrast the healthy, homeostatic cell with those experiencing aging or pathological disorders such as cancer (18–19).

Topics Summary

A. Generalized animal cell, plasma (cell) membrane

B. Movement of materials across plasma membranes

C. Cytoplasm, organelles

D. Cell inclusions, extracellular materials

E. Cell division: mitosis and cytokinesis

F. Applications to health: cells and aging

G. Cells and cancer; key medical terms associated with cells

Objectives

1. Define and list a cell's generalized parts.

2. Describe the structure and molecular organization of the plasma membrane.

3. Define diffusion, facilitated diffusion, osmosis, filtration, dialysis, active transport, phagocytosis, and pinocytosis.

4. Describe the structure and function of several modified plasma membranes.

5. Describe the chemical composition and list the functions of cytoplasm.

6. Describe two general functions of a cell nucleus.

7. Distinguish between agranular and granular endoplasmic reticulum.

8. Define the function of ribosomes.

9. Describe the role of the Golgi complex in the synthesis, storage, and secretion of glycoproteins.

10. Describe the function of mitochondria as ''powerhouses of the cell.''

11. Explain why a lysosome in a cell is called a ''suicide packet,'' and describe microtubules.

12. Describe the structure and function of centrioles in cellular reproduction.

13. Differentiate between cilia and flagella.

14. Define cell inclusion and give several examples.

15. Define extracellular material and give several examples.

16. Describe the stages and events involved in cell division.

17. Discuss the significance of cell division.

18. Describe the relationship of cancer and aging to cells.

19. Define key medical terms associated with cells.

(pages 27–28)   **A.** Generalized animal cell, plasma (cell) membrane

    1. Define:

       a. Cell

       b. Cytology

       c. Generalized animal cell

    2. List the four principal parts of a generalized animal cell.

    3. Draw and label a diagram of the plasma (cell) membrane as described by the *fluid mosaic hypothesis.*

Now complete this exercise about the membrane.

    a. The unit membrane consists of two main types of chemical compounds.

       One is _____ arranged in two parallel rows or sheets.

    b. The other membrane component is _____ . Look at the position of proteins in the diagram you drew above (and also in Figure 2–2 in your text). Explain why this hypothesis of cell membrane structure is sometimes called the "raft and iceberg" model.

c. Note the presence of pores in the membrane. What is their function?

4. What are two major functions of the plasma membrane?

5. Describe the permeability of the plasma membrane by answering these questions.
   a. Most proteins (*do? do not?*) pass through the plasma membrane. Explain.

   b. Substances that dissolve in (*lipid? water?*) can more readily pass across plasma membranes. Explain.

   c. Ions can cross the plasma membrane more easily if they have the (*same? opposite?*) charge as the membrane.
   d. Plasma membranes have special molecules that help to transport substances across membranes; these molecules are appropriately named

   _____ .

B. Movement of materials across plasma membranes   **(pages 28–33)**

1. Why must substances move across your plasma membranes in order for you to survive and maintain homeostasis?

   Give several examples of substances which must move:
   a. Into cells

   b. Out of cells

2 2. Match the processes listed below with the descriptive phrases that follow.

_____ a. Net movement of any substance (such as cocoa powder in hot milk) from region of higher concentration to region of lower concentration; membrane not required

_____ b. Same as (a) except movement across a semipermeable membrane with help of a carrier; ATP not required

_____ c. Movement of ions from region of low ion concentration to high concentration; requires ion-carrier complex, membrane, and ATP

_____ d. "Cell drinking"

_____ e. "Cell eating"

_____ f. Combination of (d) and (e)

_____ g. Net movement of water from region of high water concentration (such as 2 percent NaCl) to region of low water concentration (such as 10 percent NaCl) across semipermeable membrane; important in maintenance of normal cell size and shape

_____ h. Movement of molecules from high pressure zone to low pressure zone, for example, in response to force of blood pressure

_____ i. Movement of small molecules across a semipermeable membrane, leaving large molecules behind; the principle used in artificial kidney machines

A. Active transport
Dia. Dialysis
Dif. Diffusion
E. Endocytosis
Fac. Facilitated diffusion
Fil. Filtration
O. Osmosis
Pha. Phagocytosis
Pin. Pinocytosis

3. Contrast _active_ with _passive transport_ processes.

3 Which of the processes listed in exercise 2 above are:

a. Active

b. Passive

4. A large part of the human diet is starch. When starch is digested, it is broken down to glucose. Answer these questions related to absorption of glucose.

a. Explain why glucose cannot readily pass through plasma membranes by simple diffusion.

b. By what process does it cross plasma membranes as it is absorbed into cells lining the digestive tract?

5. Complete the following exercise about osmosis in blood.　　　　　$\boxed{4}$

    a. Human red blood cells (RBCs) contain about _____ NaCl.

        A. 2.0 percent   B. 0.85 percent   C. 0 percent (pure water)

    b. A solution that is hypertonic to RBCs contains (*more? fewer?*) solute particles and (*more? fewer?*) water molecules than blood.

    c. Which of these solutions is hypertonic to RBCs?

        A. 2.0 percent NaCl   B. 0.85 percent NaCl   C. pure water

    d. If RBCs are surrounded by hypertonic solution, water will tend to move (*into? out of?*) them, so they will (*shrink? enlarge?*).

    e. A solution that is _____ -tonic to RBCs will maintain the shape and size of the RBC. An example of such a solution

      is _____ .

    f. Which solution will cause RBCs to hemolyze?

        A. 2.0 percent NaCl   B. 0.85 percent NaCl   C. pure water

6. When blood is dialyzed in an artificial kidney, why must nutrients and certain ions be added to the blood?

7. Explain, using a diagram, how active transport occurs. Show the roles of both the carrier and ATP.

8. Define and give one example of location of each of the following modification of plasma membranes: *microvilli, stereocilia, myelin sheath.*

(pages 33–40) **C. Cytoplasm, organelles**
1. Define cytoplasm according to its chemical composition and functions.

2. The cell is compartmentalized by the presence of organelles. Of what advantage is this?

3. Study Figures 2-1 and 2-8 in your text. Then on Figure LG 2-1 label the following parts of a generalized animal cell: *plasma membrane, cytoplasm, nucleus, nuclear membrane, nucleolus, chromatin, endoplasmic reticulum (rough and smooth), ribosomes, Golgi complex, mitochondria, lysosomes,* and *centrosome.*

4. Write a sentence describing the main functions of each of these organelles:
   a. Rough ER

   b. Smooth ER

   c. Centrosome

   d. Golgi complex

   e. Microtubules

   f. Mitochondria

   g. Nucleus

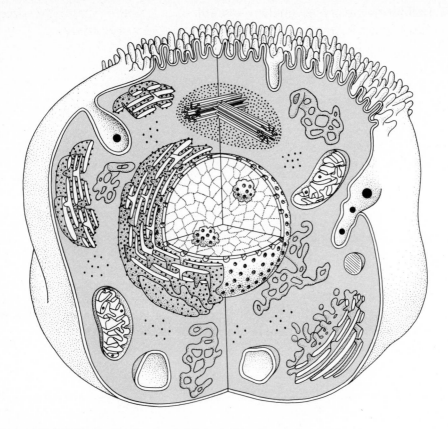

Figure LG 2-1 Generalized animal cell. Label as directed.

5. Describe the sequence of events in synthesis, secretion, and discharge of protein from the cell by numbering these five steps in correct order. 5

_____ A. Proteins accumulate in cisternae of Golgi complex.

_____ B. Proteins pass through ER to Golgi complex.

_____ C. Secretory granules move toward cell surface where they are discharged.

_____ D. Proteins are synthesized at ribosomes on rough ER.

_____ E. Vesicles containing protein pinch off of Golgi complex cisternae to form secretory granules.

6. Explain how lysosomes function in:
   a. Phagocytosis by white blood cells

   b. Bone growth and reshaping

7. Contrast *flagella* with *cilia*.

D. Cell inclusions, extracellular materials

1. Contrast *organelle* with *inclusion*.

List five examples of cell inclusions.

6   2. Name three examples of *extracellular materials*.

3. Complete the table about two major classes of matrix materials: *amorphous* and *fibrous*. Indicate to which category each of the substances listed below belongs. Then give an example of a location in the body where each type of matrix is found.

Matrix Material	Class	Location
a. Collagen		
b. Hyaluronic acid		
c. Chondroitin sulfate		
d. Elastin		

E. Cell division: mitosis and cytokinesis

1. Once you have reached adult size, do your cells continue to divide? Why or why not?

7   2. Complete this exercise about chromosomes, genes, and DNA.

a. In the nucleus of each cell are structures that contain hereditary information and also direct cell activities. These structures are known as

_____.

b. Every body cell (except mature sperm and ova) contains 46 chromosomes. Each chromosome contains about _____ genes.

c. Each gene consists of DNA arranged in repeating units known as

_____. Refer to Fig. 2-15(a) in your text.
Nucleotides are bonded to one another in DNA forming a spiral ladder

known as a _____. Sides of the ladder are composed

of alternating _____ and _____.

d. Rungs of the DNA ladders consist of molecules known as

_____. They are paired in complementary fashion, with ade-

nine (A) bonded to _____ (____), and cytosine (C)

bonded to _____ (____).

3. Cell division involves two steps: nuclear division, called _____

_____, and cytoplasmic division, called _____

_____.

4. Carefully study the phases of mitosis shown in Figure 2-16 in your text. 8
Then check your understanding of major events of each phase by doing this
matching exercise.

_____ a. Chromosomes are moved toward   A. Anaphase
           opposite poles of the cell.   I. Interphase
_____ b. Nuclear membrane disappears;   M. Metaphase
           chromatin thickens into distinct   P. Prophase
           chromosomes (chromatids).   T. Telophase
_____ c. The series of events is essentially
           the reverse of prophase; cytokin-
           esis occurs.
_____ d. Chromatids line up on the equa-
           torial plate and are attached to
           spindle fibers.
_____ e. Cell is not involved in cell divi-
           sion; it is in a metabolic phase
           which follows telophase.

F. Applications to health: cells and aging   **(pages 45–46)**

1. Briefly describe the relationship between homeostasis and aging.

2. List five generalizations that can be made about aging.

3. Note specific effects of aging on each of the following structures. Write a sentence about each.

    a. Skin

    b. Bones

    c. Muscles

    d. Brain

    e. Eyes

    f. Ears

    g. Hormone-producing (endocrine) glands

    h. Heart

    i. Blood vessels

    j. Stomach, intestine

    k. Prostate gland

    l. Kidneys

3

Tissues

Cells of similar structure and function, along with intercellular substances they produce, are organized into tissues. In this chapter you will consider a variety of specialized tissue types which make up organs of the body (Objectives 1–2). You will study structure, functions, and locations of epithelium (3–7), including that composing glands (8–10). You will contrast different kinds of connective tissue (11–16). You will see the arrangement of a number of tissue types in membranes (17–18).

Topics Summary

A. Types of tissues
B. Epithelial tissue
C. Connective tissue
D. Muscle tissue and nervous tissue
E. Membranes

Objectives

1. Define a tissue.
2. Classify the tissues of the body into four major types.
3. Describe the distinguishing characteristics of epithelial tissue.
4. Contrast the structural and functional differences of covering, lining, and glandular epithelium.
5. Compare the shape of cells and the layering arrangements of covering and lining epithelium.
6. Explain the various kinds of cell junctions that exist between epithelial cells.
7. List the structure, function, and location of simple, stratified, and pseudostratified epithelium.
8. Define a gland.
9. Distinguish between exocrine and endocrine glands.
10. Classify exocrine glands according to structural complexity and physiology.
11. Describe the distinguishing characteristics of connective tissue.
12. Contrast the structural and functional differences between embryonic and adult connective tissues.
13. Describe the ground substance, fibers, and cells that constitute connective tissue.
14. List the structure, function, and location of loose connective tissue, adipose tissue, and dense, elastic, and reticular connective tissue.
15. List the structure, function, and location of the three types of cartilage.
16. Distinguish between interstitial and appositional growth of cartilage.
17. Define an epithelial membrane.
18. List the location and function of mucous, serous, synovial, and cutaneous membranes.

Learning Activities

(page 53) **A. Types of tissues**

1. Define *tissue*.

2. Complete the table about the four main classes of tissues.

Tissue	General Functions
a. Epithelium	
b. Connective tissue	
c.	Movement
d.	Transmits nerve impulses which coordinate body activities

(pages 53–64) **B. Epithelial tissue**

1. State two general locations of epithelial tissue.

2. Write six structural characteristics of all epithelial tissue.

3. Defend or dispute this statement: "It is advantageous that all epithelium is avascular."

4. Contrast *simple* with *stratified* epithelium.

5. Define *cell junctions*, and state two general functions of these structures.

6. Contrast each of the following types of cell junctions by completing the following table.

Type	Description of Structure	Function, Location
a.		Spot welds to firmly bond cells; found in epidermis of skin.
b. Tight junctions		
c.	Contains microfilaments.	Binds cells less firmly than tight junctions.
d.		Permits intercellular passage of ions and other small molecules, e.g., to allow coordinated contraction of smooth or cardiac muscle.

28

7. Complete the table about epithelial types.

Type of Epithelium	Sketch of Tissue	Locations and Functions
a. Simple squamous		
b. Simple cuboidal		
c. Simple columnar (nonciliated)		
d. Simple columnar (ciliated)		
e. Stratified squamous		
f. Stratified cuboidal		
g. Stratified columnar		
h. Stratified transitional		
i. Pseudostratified		

8. Check your understanding of the most common types of epithelium by writing the name of the type after the phrase which describes it.   **1**

    Pseudostratified   Simple squamous
    Simple columnar   Stratified squamous
    Simple cuboidal   Stratified transitional

    a. Lines the inner surface of the stomach and intestine:

    _____

    b. Lines urinary tract, as in bladder, permitting distension:

    _____

    c. Lines mouth; present on outer surface of skin:

    _____

    d. Single layer of cube-shaped cells; found in kidney tubules and ducts of

    some glands: _____

    e. Lines air sacs of lungs where thin cells are required for diffusion of gases

    into blood: _____

    f. Not a true stratified; all cells on basement membrane, but some do not

    reach surface of tissue: _____

9. What is the function of *cilia* on cells lining the respiratory and reproductive tracts?   **2**

10. Define *keratin* and state its function.

11. What are *glands?* Why are glands studied in this section on epithelium?

12. Contast *exocrine glands* with *endocrine glunds.* (Endocrine glands will be studied further in Chapter 19.)

13. Match the types of glands with the descriptions given.   **3**

    _____ a. Tube-shaped   A. Acinar
    _____ b. Flask-shaped   C. Compound
                                      S. Simple
    _____ c. One-celled   T. Tubular
                                      U. Unicellular
    _____ d. Branched duct

    _____ e. Nonbranched duct

14. Draw a diagram of each of the following types of glands. Give an example of a location of each type.

Simple coiled tubular gland   Simple branched tubular gland

Simple branched acinar gland   Compound acinar gland

15. Complete the table contrasting functional classes of glands.

Type of Gland	How Secretion Is Released	Example
a. Apocrine		
b.	Cell accumulates secretory product in cytoplasm; cell dies; cell and its contents discharged as secretion.	
c.		Salivary glands

(pages 64–72) **C. Connective tissue**

1. Contrast connective tissue with epithelium according to location, blood supply, and amount of intercellular material.

2. What are the main functions of connective tissue?

3. The major factor that differentiates one type of connective tissue from another is (*appearance of cells? kind of intercellular substance?*).

4

4. Name several kinds of intercellular materials. (Note that *inter*cellular means "between cells" or "extracellular.")

5. Contrast loose connective tissue with dense connective tissue according to structure and location.

6. The embryonic tissue from which all other connective tissues arise is called _____. Another kind of connective tissue of the embryo, called _____, is located only in the umbilical cord.

7. How can injection of hyaluronidase along with injected drugs help lessen pain associated with the procedure?

8. Name three types of fibers in connective tissue. Write a sentence describing each.

9. Identify characteristics of each of the kinds of cells found in connective tissue by writing the name of the correct cell type after the related description. |5|

   a. Gives rise to antibodies, so helpful in defense: _____

   b. Phagocytic cell; engulfs bacteria and cleans up debris; important during infection: _____

   c. Fat cell: _____

   d. Believed to form collagenous fibers in injured tissue: _____

   e. Believed to produce heparin, an anticoagulant, as well as histamine, which dilates blood vessels:

   _____

   Adipose
   Fibroblast
   Macrophage
   Mast cell
   Plasma cell

10. Match the common types of dense connective tissue with descriptions given.

⟨6⟩

_____ a. Connects muscles to bones

_____ b. Holds bones together at joints

_____ c. Flat band or sheet of tissue connecting muscles to each other or to bones

_____ d. Sheet of connective tissue wrapped around muscle bundles, holding them in place

A. Aponeurosis
D. Deep fascia
L. Ligament
T. Tendon

11. In general, cartilage can endure (*more? less?*) stress than the connective tissues you have studied so far.

⟨7⟩ 12. Match the types of cartilage with descriptions given.

_____ a. Found where strength and rigidity are needed, as in discs between vertebrae and in symphysis pubis

_____ b. White, glossy cartilage covering ends of bones (articular), ends of ribs (costal), and giving strength to nose, larynx, and trachea

_____ c. Provides strength and flexibility, as in external part of ear.

E. Elastic
F. Fibrous
H. Hyaline

13. Contrast *interstitial* (endogenous) growth of cartilage with *appositional* (exogenous) growth.

14. Review connective tissue types in Exhibit 3-2 of your text. Then examine the diagrams in Figure LG 3-1 and label different cell types and structures characteristic of each tissue type.

(page 72) D. Muscle tissue and nervous tissue

1. These tissues are more (*specific? generalized?*) in their structure and functions than epithelium or connective tissue. (These will be studied in Chapters 9 and 14.)

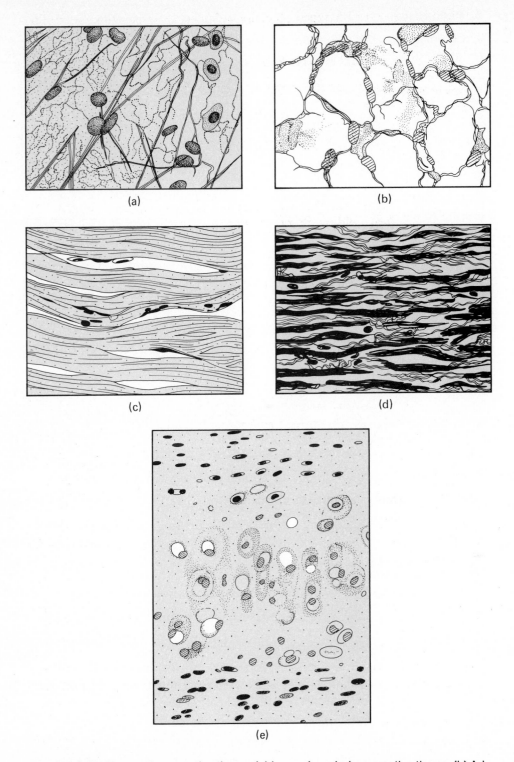

Figure LG 3-1 Types of connective tissue. (a) Loose (areolar) connective tissue. (b) Adipose. (c) Dense (collagenous). (d) Elastic. (e) Hyaline cartilage. Label as directed.

1. Complete the table about three types of membranes.

Type of Membrane	Location	Example of Specific Location	Function(s)
a.	Lines body cavities leading to exterior		
b. Serous			Allows organs to glide easily over each other
c.		Lines knee and hip joints	

8 2. Check your understanding of membrane types by doing this exercise.

a. The serous membrane covering the heart is known as the
_____ , whereas that covering the lungs is called the
_____ . The serous membrane over abdominal
organs is the _____ .

b. The portion of serous membranes that covers organs (viscera) is called
the _____ layer; that portion lining the cavity is
named the _____ layer.

c. The _____ layer of a mucous membrane binds epi-
thelium to underlying muscle and serves as a route for oxygen and nutrients to reach the avascular epithelial layer of the membrane.

d. Another name for skin is _____ membrane.

Answers to Numbered Questions in Learning Activities

1 (a) Simple columnar. (b) Stratified transitional. (c) Stratified squamous.
(d) Simple cuboidal. (e) Simple squamous. (f) Pseudostratified.

2 Cilia wave in unison to move mucus and foreign particles upward, away
from the lungs and toward the throat where they can be swallowed. Cilia
propel egg (ovum) from the ovary toward the uterus.

4

The Integumentary System

In this chapter you will see how cells, tissues, and organs are organized into a system—the integumentary system. You will learn about one of the largest organs in the body, the skin, including its functions and layers, as well as the basis for skin color (Objectives 1–5). You will examine special organs of the integument: hair, glands, and nails (6–8). Then you will learn about disorders and treatment of skin with special emphasis on burns (9–11). You will also define key medical terms associated with the integument (12).

Topics Summary

A. Skin
B. Epidermal derivatives
C. Applications to health; key medical terms

Objectives

1. Define the skin as the organ of the integumentary system.
2. Explain how the skin is structurally divided into epidermis and dermis.
3. List the various layers of the epidermis and describe their functions.
4. Explain the basis for skin color.
5. Describe the composition and function of the dermis.
6. Describe the development, distribution, structure, color, and replacement of hair.
7. Compare the structure, distribution, and functions of sebaceous and sudoriferous glands.
8. List the parts of a nail and describe their composition.
9. Describe the causes, effects, and treatment (where applicable) for the following skin disorders: acne, impetigo, systemic lupus erythematosus, psoriasis, decubitus, warts, cold sores, sunburn, and skin cancer.
10. Classify burns into first, second, and third degrees.
11. Define the "rule of nines" for estimating the extent of a burn.
12. Define key medical terms associated with the integumentary system.

Learning Activities

(pages 76–81)   **A.** Skin

1. Name the structures included in the integumentary system.

2. Describe some important aspects of skin by doing this exercise.
   a. Why is skin considered an organ?

   b. The skin is one of the largest organs in the body. It covers a surface area of 2 sq m, which is equivalent to about 21 sq ft or _____ sq inches.
   c. Draw a line 3 mm long to show the thickness of the thickest skin in the body. Where might such skin be found?

   d. Try to draw a line 0.05 mm long, the thickness of the thinnest skin in the body. (Can you?) Where might such skin be located?

   e. Skin may be one of the most underestimated organs in the body. What functions does your skin perform while it is "just lying there" covering your body? (Can you list eight functions?)

3. Answer these questions about the two portions of skin. Label them on Figure LG 4-1.

   <u>1</u>

   a. The outer layer is named the _____. It is composed of (*connective tissue? epithelium?*).

   b. The inner portion of skin, called the _____, is made of (*connective tissue? epithelium?*). The dermis is (*thicker? thinner?*) than the epidermis.

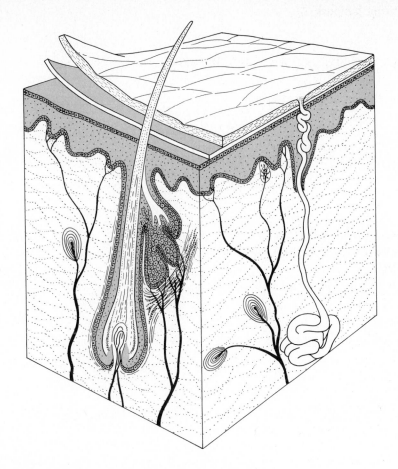

Figure LG 4-1 Structure of the skin. Label as directed.

4. The tissue underlying skin is called *subcutaneous,* meaning _____

   _____ . This layer is also called _____ .

   It consists of two types of tissue, _____ and

   _____ . What functions does subcutaneous tissue serve?

5. The layers of the epidermis are listed below from deepest to most superficial. Describe the structure and functions of each layer. Label each layer on
   Figure LG 4-1.

   a. Stratum basale

   b. Stratum spinosum

   c. Stratum granulosum

d. Stratum lucidum

e. Stratum corneum

6. Explain how skin is waterproofed in order to prevent excessive fluid loss.

7. Contrast structural differences among skins of different colors.
   a. Black (Negro)

   b. Pink (Caucasian)

   c. Yellow (Oriental)

8. Describe the roles of the following in skin color: *melanocytes, tyrosinase, UV radiation, melanin,* and *MSH.*

9. Define:
   a. Albinism

   b. Melanoma

10. Compare the dermis with the epidermis in terms of structure and function. Distinguish *papillary* from *reticular* regions of dermis.

11. What are *striae?* Under what circumstances are they most likely to develop?

12. Describe *epidermal ridges.* Explain how they determine your fingerprint pattern.

13. Define *lines of cleavage* (*Langer's lines*). State their significance in surgery.

14. Explain the role of skin in temperature regulation.

B. Epidermal derivatives
(pages 81–86)

1. What are the functions of hair?

2. Explain how hair develops. Include the terms *lanugo, vellus* (*fleece*), and *terminal hairs* in your description.

3. Which parts of the body lack hairs?

4. On Figure LG 4-1 label the following parts of a hair and its associated structures: *hair shaft, root, bulb, internal root sheath, external root sheath,* and *connective tissue papilla.*

2   5. Answer these questions about growth of hair.
   a. The part of the follicle responsible for growth of a new hair is the

   _____ .

   b. The rate of growth of hair is about 1 mm every _____ days, or about 1

   cm (0.4 inch) every _____ days.

6. Describe the function of these structures that are associated with hairs. Label the structures on Figure LG 4-1.
   a. Arrector pili muscles

   b. Sebaceous glands

7. Explain what seems to account for change of dark hair color to gray or white hair.

3   8. Indicate whether the following descriptions refer to *sebaceous* or *sudoriferous* glands.

   a. Sweat glands: *Sudoriferous*
   b. Most abundant in palms, soles, forehead, and axilla: *Sudoriferous*
   c. Simple branched acinar glands leading directly to hair follicle: *Sebaceous*

   d. Keep hair and skin from drying out: _____

9. How are "blackheads" related to *sebum?*

10. Look at one of your own nails and Figure 4-7. Identify these parts of your nail: *free edge, nail body, lunula, eponychium (cuticle), hyponychium, nail bed, nail groove,* and *nail fold.*

11. Why does the nail body appear pink, yet the lunula and free edge appear white?

12. How do nails grow?

C. Applications to health; key medical terms (pages 86–90)

1. Write a description of these disorders. Include causes, parts of the body most often affected, and groups of people (by age or other factors) most often afflicted.
   a. Acne vulgaris

   b. Decubitus ulcers

   c. Skin cancer

2. Match the name of the disorder with the description given. $\boxed{4}$

   _____ a. Bedsores

   _____ b. An autoimmune disease with many symptoms, including "butterfly rash"

   _____ c. Staphylococcal or streptococcal infection which may become epidemic in nurseries

   _____ d. Inflammation of sebaceous glands especially in chin area, occurs among women in their 20s

   _____ e. One kind of skin cancer

   _____ f. Chronic disease of 6 to 8 million people in the United States, characterized by reddish plaques or papules, most severe among ages 10–50

   A. Acne cosmetica
   D. Decubitus
   I. Impetigo
   L. Lupus or SLE
   M. Melanoma
   P. Psoriasis

3. Contrast *systemic* effects with *local* effects of burns.

4. What reasons would you give if you were advising a person against constant overexposure to sun?

5. What is *sunstroke?*

6. Identify characteristics of the three different classes of burns by completing this matching exercise.

5

___S___ a. Damage is restricted to epidermis and parts of dermis. Regeneration of epithelium is possible.

F. First-degree
S. Second-degree
T. Third-degree

___T___ b. Both epidermis and dermis are destroyed. Regeneration is slow and usually leads to scars.

___F___ c. Only epidermal tissue is destroyed. Symptoms are redness, pain, and edema.

7. Write the percentages of the total body surface that are found in each of the areas listed and that are used to estimate extent of burns according to the "rule of nines." (Hint: The percentages should add up to 100.)

6

___9%___ a. Head and neck

___18%___ b. Anterior portion of trunk

___18%___ c. Posterior portion of trunk

___9%___ d. The entire right leg

___9%___ e. The entire left leg

___18%___ f. The two arms together

___1%___ g. The perineum

5

Osseous Tissue

In the next five chapters you will consider the parts of the body that provide support and the ability to move—skeleton, joints, and muscles. In Chapter 5 you will take a brief look at the components and functions of the skeletal system (Objective 1). You will examine the gross and microscopic structure of bone tissue (2–4) and the manner in which bones form and continue to grow (5–9). You will study some applications to health and medical terminology related to the skeletal system, with particular emphasis on fractures (10–14).

Topics Summary

A. Functions
B. Histology
C. Ossification
D. Applications to health; key medical terms

Objectives

1. Identify the components of the skeletal system.
2. Explain the gross features of a long bone.
3. Describe the histological features of dense bone tissue.
4. Compare the histological characteristics of spongy and dense bone tissue.
5. Contrast the steps involved in intramembranous and endochondral ossification.
6. Identify the zones and growth pattern of the epiphyseal plate.
7. Describe the processes of bone construction and destruction involved in bone replacement.
8. Describe the blood and nerve supply of bone tissue.
9. Discuss the conditions necessary for normal bone growth.
10. Define rickets and osteomalacia as vitamin deficiency disorders.
11. Contrast the causes and clinical symptoms associated with osteoporosis, Paget's disease, and osteomyelitis.
12. Define a fracture and describe several common kinds of fractures.
13. Describe the sequence of events involved in fracture repair.
14. Define key medical terms associated with the skeletal system.

(page 93) **A.** Functions

1. List and discuss briefly five functions of the skeletal system.

2. What other systems of the body depend on a healthy skeletal system? Explain why in each case.

[1]

(pages 93–97) **B.** Histology

[2]

1. In what way is intercellular substance of bone unique among connective tissues?

2. On Figure LG 5-1 label the following structures: *diaphysis, epiphysis, articular cartilage, periosteum, medullary (marrow) cavity, endosteum,* and areas of *compact bone* and *spongy bone.*

3. Contrast according to location and function:
   a. Fibrous periosteum

   b. Osteogenic periosteum

4. What are functions of each of these structures?
   a. Articular cartilage

   b. Medullary (marrow) cavity

   c. Endosteum

Figure LG 5-1 Diagram of a long bone that has been partially sectioned lengthwise. Label as directed.

5. Describe some functional advantages provided by the following aspects of long bones.
   a. Long bones are hollow cylinders, not solid bone.

   b. Much of bone is spongy, porous bone, but a layer of compact bone is present on the outside of bones.

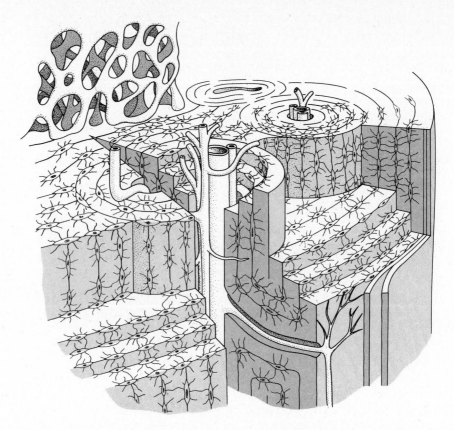

Figure LG 5-2 Haversian systems of compact bone. Label as directed.

c. Ends (epiphyses) of bones are bulbous, while shafts (diaphyses) are much narrower in diameter.

6. On Figure LG 5-2 label these parts of a Haversian system (osteon): *lamellae, lacunae, osteocytes, canaliculi,* and *blood vessels in Haversian canals* and *Volkmann's canals.*

7. On Figure LG 5-2, color blood vessels red and osteocytes blue. Note that these colored parts are the only living tissue in bone. Typical of all connective tissues, bone consists mainly of _____ .

8. Check your understanding of bone tissue by writing a term that fits each description. Note that most of these structures are shown in Figure LG 5-2.

   ⎣4⎦

   a. Spaces containing bone cells: _____

   b. Bone cells: _____

   c. Concentrically arranged calcified layers of bone: _____
   _____

   d. Small (microscopic) canals running longitudinally through bone and containing blood vessels: _____

   e. Large (macroscopic) cavity filled with yellow marrow: _____

f. Minute canals, radially arranged between lacunae: _____

g. Horizontal canals carrying blood vessels from periosteum to Haversian systems and marrow: _____

C. Ossification (pages 97–103)

1. What is mesenchyme? What does it have to do with the skeleton?

2. The two main kinds of tissue that compose the "skeleton" of a developing embryo or fetus are _____ and _____ _____ .

   5

3. Which bones of the body form by the process of intramembranous ossification?

4. Describe the process of intramembranous ossification. Include these terms in your paragraph: *mesenchymal cells, osteoblasts, center of ossification, collagenous fibers, calcium salts, trabeculae, spongy bone,* and *osteocytes.*

5. Write a sentence about each of the steps of endochondral ossification using these key terms in order:
   a. Perichondrium

   b. Collar of bone

   c. Primary ossification center

d. Calcification

e. Death of cartilage cells

f. Spaces, marrow

g. Secondary ossification centers

h. Articular cartilage

i. Epiphyseal plate

6. Match the names of the layers of the epiphyseal plate with the correct descriptions.

_____ a. Cells do not function in bone growth, but anchor epiphyseal plate to bone of epiphysis.

_____ b. New cartilage cells are made by mitosis.

_____ c. Cartilage cells mature, surround themselves with calcium salts, then die due to these salts.

_____ d. Dead cartilage cells and broken up matrix are replaced by bone-forming cells.

C. Zone of calcified matrix
H. Zone of hypertrophic cartilage
P. Zone of proliferating cartilage
R. Zone of reserve cartilage

7. Complete this summary statement about bone growth at the epiphyseal plate.

Cartilage cells multiply on the (*epiphysis? diaphysis?*) side of the epiphyseal plate, providing temporary new tissue. But cartilage cells then die and are replaced by bone cells on the (*epiphysis? diaphysis?*) side of the epiphyseal plate.

8. Contrast the *epiphyseal plate* with the *epiphyseal line.*

9. Explain how bones grow in diameter. Contrast the roles of *osteoblasts* and *osteoclasts.*

10. Look at the roentgenograms in Figure 5-7 in your text. In these films cartilage appears (*darker? lighter?*) than bone. Explain why this is so. ⬛8

11. Defend or dispute this statement: "Once a bone, such as your thighbone, is formed, the bone tissue is never replaced unless the bone is broken."

D. **Applications to health; key medical terms** (pages 103–109)

1. Describe the role of vitamin D in normal bone growth and development.

2. Check your understanding of bone disorders by matching the terms with related descriptions. ⬛9

   a. Benign bone tumor: _____   Osteosarcoma
                                                Osteoma
   b. Malignant bone tumor: _____   Osteomyelitis
   c. Bone infection, for example, caused by   Osteoporosis
      staphylococcus: _____   Paget's disease
   d. Extra bone produced in some areas; bone   Rickets
      destroyed in other areas; usually affects

      persons over 50 years: _____
   e. Caused by vitamin D deficiency which
      prevents normal calcium absorption; soft

      bones: _____
   f. Associated with decreased sex hormones,
      especially in women after menopause;

      decreased bone mass: _____

3. Contrast the terms related to types of fractures in each pair.
   a. Simple/compound

b. Partial/complete

c. Pott's/Colles'

4. Without referring to figure captions, identify each of the types of fractures shown in Figure 5-9 in your text.
5. What is the major difference between these two methods of setting a fracture: *closed reduction* and *open reduction?*

6. Describe the steps involved in repair of a fracture.

Answers to Numbered Questions in Learning Activities

1.   Essentially all do. For example, muscles need intact bones for movement to occur; bones are site of blood formation; bones provide protection for viscera of nervous, digestive, urinary, reproductive, cardiovascular, respiratory, and endocrine systems; broken bones can injure integument.

2.   It contains abundant mineral salts.

3.   (a) Skeleton is lighter than it would be if it were solid bone. (b) Though skeleton is light, it is strong. (c) Joints are stable with broad area for muscle attachments.

4.   (a) Lacunae. (b) Osteocytes. (c) Lamellae. (d) Haversian canals. (e) Medullary (marrow) cavity. (f) Canaliculi. (g) Volkmann's canals.

5.   Hyaline cartilage, fibrous membranes.

6.   (a) R. (b) P. (c) H. (d) C.

6

The Skeletal System: The Axial Skeleton

In this chapter you will see how the construction of a bone—its size, shape, and markings—relates to its functions (Objectives 1–3). You will distinguish the two principal classifications of bones, axial and appendicular, and their components (4). You will then examine bones of the axial skeleton: skull (5–8), vertebral column (9–10), and thorax (11). Finally, you will study some applications to health associated with the axial skeleton (12).

Topics Summary

A. Types of bones
B. Surface markings
C. Divisions of the skeletal system
D. Skull
E. Hyoid bone
F. Vertebral column
G. Thorax
H. Applications to health

Objectives

1. Define the four principal types of bones in the skeleton.
2. Describe the various markings on the surfaces of bones.
3. Relate the structure of the marking to its function.
4. List the components of the axial and appendicular skeletons.
5. Identify the bones of the skull and the major markings associated with each.
6. Identify the sutures and fontanels of the skull.
7. Identify the paranasal sinuses of the skull in projection diagrams and roentgenograms.
8. Identify the principal foramina of the skull.
9. Identify the bones of the vertebral column and their principal markings.
10. List the defining characteristics and curves of each region of the vertebral column.
11. Identify the bones of the thorax and their principal markings.
12. Contrast microcephalus, hydrocephalus, slipped disc, curvatures, spina bifida, and fractures of the vertebral column as disorders associated with the skeletal system.

Learning Activities

(pages 112–113) **A. Types of bones**

1. Complete the table about the four major and two minor types of bones.

Type of Bone	Structural Features	Examples
a. Long	Slightly curved to absorb stress better	
b.		Wrist, ankle bones
c.	Composed of two thin plates of bone	
d. Irregular		
e. Wormian		
f.	Small bones in tendons	

(page 113) **B. Surface markings**

1. In general, what is the purpose of surface markings of bones?

2. Contrast the bone markings in each of the following pairs.
   a. Tubercle/tuberosity

   b. Crest/line

   c. Fossa/foramen

   d. Condyle/epicondyle

3. Match the general descriptions of markings with the examples of specific bone markings. Take particular note of italicized terms. [1]

 a. Air-filled cavity within a bone, connected to nasal cavity: _____

   _____

 b. Narrow, cleftlike opening between adjacent parts of bone; passageway for blood vessels and nerves: _____

   _____

 c. Rounded hole, passageway for blood vessels and nerves: _____

   _____

 d. Tubelike passageway through bone: _____

 e. Large, rounded projection above constricted neck: _____

   _____

 f. Large, blunt projection on the femur: _____

 g. Sharp, slender projection: _____

   _____

 h. Smooth, flat surface: _____

   _____

Articular *facet* on vertebra
External auditory *meatus*
Optic *foramen*
Maxillary *sinus*
Greater *trochanter*
Styloid *process*
Superior orbital *fissure*
Head of humerus

C. Divisions of the skeletal system (pages 113–114)

1. Describe the bones in the two principal divisions of the skeletal system by completing this exercise. [2]

 a. Bones that lie around the axis of the body are included in the (*axial? appendicular?*) skeleton.
 b. The axial skeleton includes the following groups of bones. Indicate how many bones are in each category.

   _____ Skull (cranium, face)   _____ Vertebrae

   _____ Earbones   _____ Sternum

   _____ Hyoid   _____ Ribs

 c. The total number of bones in the axial skeleton is _____.
 d. The appendicular skeleton consists of bones in which parts of the body?

 e. Write the number of bones in each category. Note that you are counting bones on one side of the body only.

   _____ Left shoulder girdle   _____ Left hipbone

   _____ Left upper extremity (arm, forearm, wrist, hand)   _____ Left lower extremity (thigh, kneecap, leg, foot)

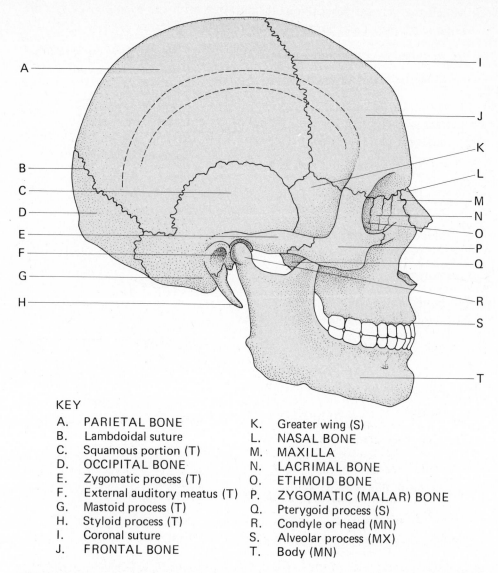

KEY

A. PARIETAL BONE
B. Lambdoidal suture
C. Squamous portion (T)
D. OCCIPITAL BONE
E. Zygomatic process (T)
F. External auditory meatus (T)
G. Mastoid process (T)
H. Styloid process (T)
I. Coronal suture
J. FRONTAL BONE

K. Greater wing (S)
L. NASAL BONE
M. MAXILLA
N. LACRIMAL BONE
O. ETHMOID BONE
P. ZYGOMATIC (MALAR) BONE
Q. Pterygoid process (S)
R. Condyle or head (MN)
S. Alveolar process (MX)
T. Body (MN)

Figure LG 6-1 Right lateral view of the skull. Abbreviations in key: MN, mandible; MX, maxilla; S, sphenoid; T, temporal. Color as indicated.

f. The total number of bones in both (right and left) upper extremities and shoulder girdles is _____.

g. The total number of bones in both (right and left) hipbones and lower extremities is _____.

h. There are _____ bones in the appendicular skeleton.

i. In the entire human body there are _____ bones.

(pages 115–128) **D. Skull**

1. Try to locate the major bones of the skull by using these aids: text Figures 6-2, 6-4, and 6-5; a mirror to examine your own facial contours, and a skull specimen (if available). At this time, do not concentrate on specific markings of bones; rather, try to identify bones by name and position relative to other bones of the skull. Be sure to find all skull bones in the following list. Color each bone on Figures LG 6-1 and LG 6-2. Use colors indicated in parentheses. (Code letters for each bone will be used in exercises below.)

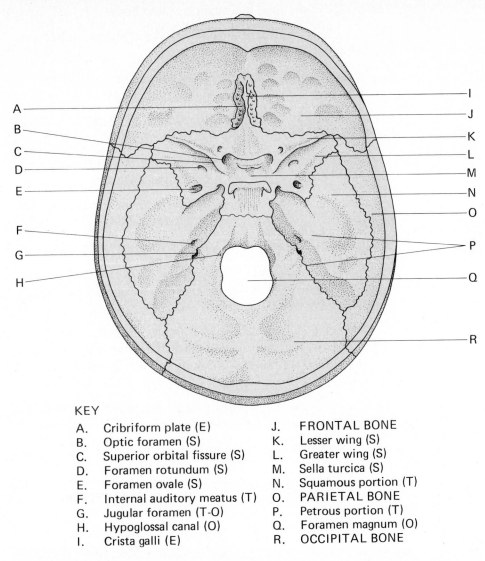

KEY

A.	Cribriform plate (E)	J.	FRONTAL BONE
B.	Optic foramen (S)	K.	Lesser wing (S)
C.	Superior orbital fissure (S)	L.	Greater wing (S)
D.	Foramen rotundum (S)	M.	Sella turcica (S)
E.	Foramen ovale (S)	N.	Squamous portion (T)
F.	Internal auditory meatus (T)	O.	PARIETAL BONE
G.	Jugular foramen (T-O)	P.	Petrous portion (T)
H.	Hypoglossal canal (O)	Q.	Foramen magnum (O)
I.	Crista galli (E)	R.	OCCIPITAL BONE

Figure LG 6-2 Floor of cranium. Abbreviations in key: E, ethmoid; O, occipital; S, sphenoid; T, temporal. Color as indicated.

List of Skull Bones

E. Ethmoid (blue)
F. Frontal (red)
I. Inferior nasal concha (purple)
L. Lacrimal (yellow)
Man. Mandible (blue)
Max. Maxilla (green)
N. Nasal (yellow)

O. Occipital (red)
Pal. Palatine (orange)
Par. Parietal (blue)
S. Sphenoid (yellow)
T. Temporal (orange)
V. Vomer (red)
Z. Zygomatic (purple)

2. Check your understanding of location and general functions of each skull bone by matching the correct bones with the following descriptions. Use the code letters from the list above.

      3

_____ a. This bone forms the lower jaw, including the chin.

_____ b. These are the cheek bones; they also form lateral walls of the orbit of the eye.

_____ c. Tears pass through tiny foramina in these bones; they are the smallest bones in the face.

_____ d. The bridge of the nose is formed by these bones.

_____ e. Organs of hearing (internal part of ears) are located in and protected by these bones.

_____ f. This bone sits directly over the spinal column; it contains the hole through which the spinal cord connects to the brain.

_____ g. The name means "wall." The bones form most of the roof and much of the side walls of the skull.

_____ h. These bones form most of the roof of the mouth (hard palate) and contain the sockets into which upper teeth are set.

_____ i. L-shaped bones form the posterior parts of the hard palate and nose.

_____ j. Commonly called the forehead, it provides protection for the anterior portion of the brain.

_____ k. A light spongy bone, it forms much of the roof and internal structure of the nose.

_____ l. It serves as a "keystone," since it binds together many of the other bones of the skull. It is shaped like a bat, with the wings forming part of the sides of the skull and the legs at the back of the nose.

_____ m. This bone forms the inferior part of the septum dividing the nose into two nostrils.

_____ n. Two delicate bones form the lower parts of the side walls of the nose.

3. What is the main function of the _cranium?_

| 4 | Which bones are considered parts of the cranium, rather than the face? Write code letters from the list of skull bones. |

4. Which six of the bones of the skull are unpaired (that is, there is only one

| 5 | bone of the name)? |

5. What are sutures?

Identify locations of each of these sutures.
a. Coronal

b. Sagittal

c. Lambdoidal

d. Squamosal

6. Of what functional advantage are "soft spots" of the skull during delivery of a baby?

7. Try to identify all of the structures on Figures LG 6-1 and LG 6-2. Take particular note of markings and recognize what bones they are parts of (indicated in parentheses in the key).

8. Complete the table describing major markings of the skull.

Marking	Bone	Function
a. Greater wings	Sphenoid	
b. External auditory meatus		
c.		Site of pituitary gland
d. Petrous portion		
e.		Largest hole in skull; passageway for spinal cord
f.		Passageway into skull for carotid artery
g. Occipital condyles		
h.		Site of only air sinuses that do not drain into nose
i. Pterygoid processes		
j.		Bony sockets for teeth

9. The 12 pairs of nerves attached to the brain are called cranial nerves. Holes in the skull permit passage of these nerves to and from the brain. These nerves are numbered according to the order in which they attach to the brain (and leave the cranium) from I (most anterior) to XII (most posterior). To help you visualize their sequence, foramina for cranial nerves are labeled in order on the left side of Figure LG 6-2. Complete the table summarizing these foramina. The first one is done for you.

6

Number and Name of Cranial Nerve	Location of Opening for Nerve
a. I Olfactory	Cribriform plate of ethmoid bone
b. II Optic	
c. III Oculomotor IV Trochlear V Trigeminal (ophthalmic branch) VI Abducens	
d. V Trigeminal (maxillary branch)	
e. V Trigeminal (mandibular branch)	
f. VII Facial VIII Vestibulocochlear	
g. IX Glossopharyngeal X Vagus XI Accessory	
h. XII Hypoglossal	

10. List functions of paranasal sinuses.

7

11. Name bones that contain paranasal sinuses. (Practice identifying their locations the next time you have a sinus headache!)

12. What structures form the septum of the nose?

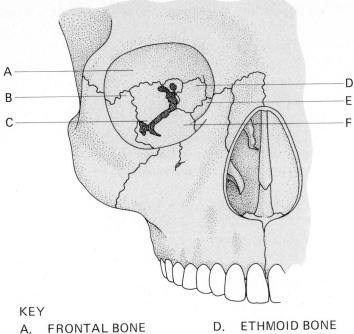

KEY

A. FRONTAL BONE
B. ZYGOMATIC BONE
C. SPHENOID BONE

D. ETHMOID BONE
E. LACRIMAL BONE
F. MAXILLA BONE

Figure LG 6-3 Bones of the right orbit. Color as indicated.

13. A good way to test your ability to visualize locations of many important skull bones is to try to identify bones that form the orbit of the eye. See if you can determine the correct positions of the six major bones comprising the orbit shown in Figure LG 6-3. Using the color code in the list of skull bones, p. LG 63, color each bone.

E. Hyoid bone (page 127)

1. In what way is the hyoid bone unique among the bones of the axial skeleton?

2. Describe the location and functions of the hyoid bone.

F. Vertebral column (pages 129–137)

1. List several functions of the vertebral column.

2. Write the names of each type of vertebra from superior to inferior. Indicate the number of each type. Identify the location of each region on yourself.

8

3. Which regions of the vertebral column are normally arranged with an anteriorly concave curve?

    C. Cervical   L. Lumbar   S. Sacral   T. Thoracic

4. On Figure LG 6-4 label the following structures: *body, pedicle, lamina, transverse process, spinous process, vertebral foramen,* and *superior articulating process.*

9

5. Choose the terms that fit the descriptions of parts of vertebrae. (Not all terms will be used.)

a. Anterior portion of vertebral arch:

    _____

b. Drum-shaped portion designed to

    bear weight: _____

c. Opening through which spinal cord

    passes: _____

d. Forms joint with vertebra be-

    low: _____

e. Space between two vertebrae where

    spinal nerves exit: _____

f. Most posterior part of verte-

    bra: _____

Body
Inferior articulating process
Intervertebral foramen
Lamina
Pedicle
Spinous process
Superior articulating process
Transverse process
Vertebral foramen

10

6. Identify distinctive features of vertebrae in each region.

    C. Cervical   Co. Coccyx   L. Lumbar   S. Sacral   T. Thoracic

_____ a. Small body, foramina for vertebral blood vessels in transverse processes

_____ b. Atlas and axis

_____ c. Five in the adult

_____ d. Only vertebrae that articulate with ribs

_____ e. Massive body, blunt spinous process

_____ f. Long spinous processes that point inferiorly

_____ g. Most inferior part of vertebral column

_____ h. Articulate with the two hipbones

7. Contrast the *atlas* and *axis* according to location, structure, and function.

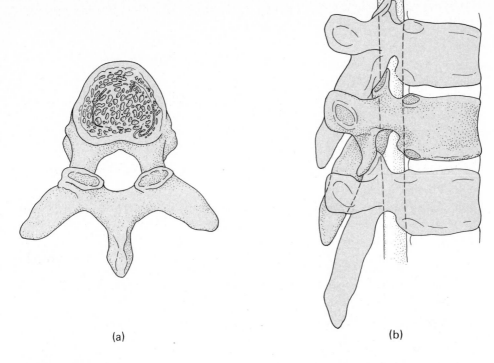

(a)

(b)

Figure LG 6-4 A typical vertebra. (a) Superior view. (b) Right lateral view. Label as directed.

8. Examine a skeleton or Figure 6-18 in your text to see how ribs articulate with thoracic vertebrae. Describe differences in the manner of articulation along the vertebral column which account for differences in facets and demifacets on vertebrae.

G. Thorax

(pages 137–139)

1. Name the structures that compose the thorax. (Why is it called a cage, the "thoracic cage"?)

2. Listed below are the three portions of the sternum. Tell what structures (if any) articulate with each part.
   a. Manubrium

b. Body

c. Xiphoid process

11   3. Complete this exercise about ribs.

a. There are a total of _____ ribs (_____ pairs) in the human skeleton.
b. Ribs slant in such a way that the anterior portion of the rib is (*superior? inferior?*) to the posterior end of the rib.

c. Posteriorly, all ribs articulate with _____. Ribs 1–10 also pass (*anterior? posterior?*) to and articulate with transverse processes of vertebrae. (Check this on a skeleton or Figure 6-18(c) of the text.) Of what functional advantage is such an arrangement?

d. Anteriorly, ribs numbered _____ to _____ attach to the sternum directly by means of strips of hyaline cartilage, called _____ cartilage. These ribs are called (*true? false?*) ribs.

e. Ribs numbered 8 to 10 are called _____ ribs. Do they attach to the sternum? (*Yes? No?*) If so, in what manner?

f. Ribs numbered _____ to _____ are called "floating ribs." Why?

g. The head and neck of a rib are located at the (*anterior? posterior?*) end.
h. What function is served by the *costal groove?*

i. What occupies *intercostal spaces?*

(pages 139–141)   **H. Applications to health**

1. Contrast *microcephalus* with *hydrocephalus.* Explain how hydrocephalus may be treated.

The Skeletal System: The Appendicular Skeleton

In the previous chapter you learned about the bones comprising the axial skeleton. In this chapter you will study the remaining bones—those in the appendicular skeleton. You will identify bones of the shoulder girdles (Objective 1), upper extremities (2), pelvic girdle (3), and lower extremities (4–5). You will also consider structural differences between the male and female skeletons (6).

Topics Summary

A. Shoulder girdles
B. Upper extremities
C. Pelvic girdle
D. Lower extremities
E. Male and female skeletons

Objectives

1. Identify the bones of the shoulder girdle and their major markings.
2. Identify the upper extremity, its component bones, and their markings.
3. Identify the components of the pelvic girdle and their principal markings.
4. Identify the lower extremity, its component bones, and their markings.
5. Define the structural features and importance of the arches of the foot.
6. Compare the principal structural differences between male and female skeletons, especially those that pertain to the pelvis.

76   **Learning Activities**

(page 144) **A. Shoulder girdles**

1. Name the bones that comprise the shoulder girdle. Locate these bones on yourself or a partner. State whether they articulate with vertebrae or ribs.

   1. Clavicle — ribs but sternum & scap.

   2. Scapula — no vert.
   art. with clav & scap.

2. How does the medial end of the clavicle differ in shape from the lateral end?

   medial is round & art with sternum.

   lateral is broad & flat art with acromian of scap.

3. Study a scapula carefully, using a skeleton or Figure 7-3 in your text. Then match the following markings with the descriptions given.

   1

   S a. Sharp ridge on the posterior surface   A. Axillary border
   I b. Depression inferior to the spine; location of infraspinatus muscle   C. Coracoid process
       I. Infraspinatus fossa
   M c. Edge closest to the vertebral column   M. Medial border
       S. Spine
   A d. Thick edge closest to the arm
   C e. Projection on the anterior surface; used for muscular attachment

   2

4. Write the names of bones that articulate with these areas of the scapula.
   a. Acromion process   *clavical*

   b. Glenoid cavity   *humerus head.*

(pages 144–151) **B. Upper extremities**

1. List the names of the bones of the upper extremity, from proximal to distal.

   humerous, ulna, radius, carpals, meta carpals, phalanges.

2. After carefully examining Figures 7-4 and 7-5 in your text, as well as bone specimens (if available), complete this exercise. For each marking write the name of the bone in which it is found, its location on the bone, and its function. The first one is done for you.

Marking	Bone	Location	Function
a. Head	Humerus	Proximal end	Articulates with glenoid cavity of scapula
b. Intertubercular sulcus	*Humerus*	*proximal end.*	*Biceptal grove.*

Marking	Bone	Location	Function
c. Deltoid tuberosity	humerous	middle of shaft	attachment for deltoid
d. Trochlea	humerous		articulates with ulna
e. Trochlear (semilunar) notch	humerous	Between olecranon & coronoid	trochlea of humerus fits in notch
f. Coronoid process	humerous	distal of humerous	holds trochea of humerous
g. Coronoid fossa			houses ulna
h. Olecranon process	~~pro~~ ulna	proximal of ulna	forms elbow
i. Olecranon fossa	humerous	distal end of humerous	receives olecranon
j. Radial notch	ulna	proximal of ulna	houses radius
k. Radial tuberosity	radius	proximal of radius	biseps attachment
l. Radial fossa			
m. Capitulum	~~distal of~~ humerous	distal of humerus	articulate with head of radius

articutates with ulna

3. Circle all of the TRUE statements about the human arm and forearm in the anatomical position.

  3

  (A.) The radius and ulna are parallel to each other.
   B. The radius and ulna are crossed.
  (C.) The radius is lateral to the ulna.

4. Answer these questions about wrist bones.

  a. Wrist bones are called ___carpals.___ . There are (5? 7? (8?) 14?) of them in each wrist.

  b. The wrist bone most subject to fracture is the bone named ___Scaphoid___, located just distal to the (radius? ulna?).

5. Trace an outline of your hand. Draw in and label all bones.

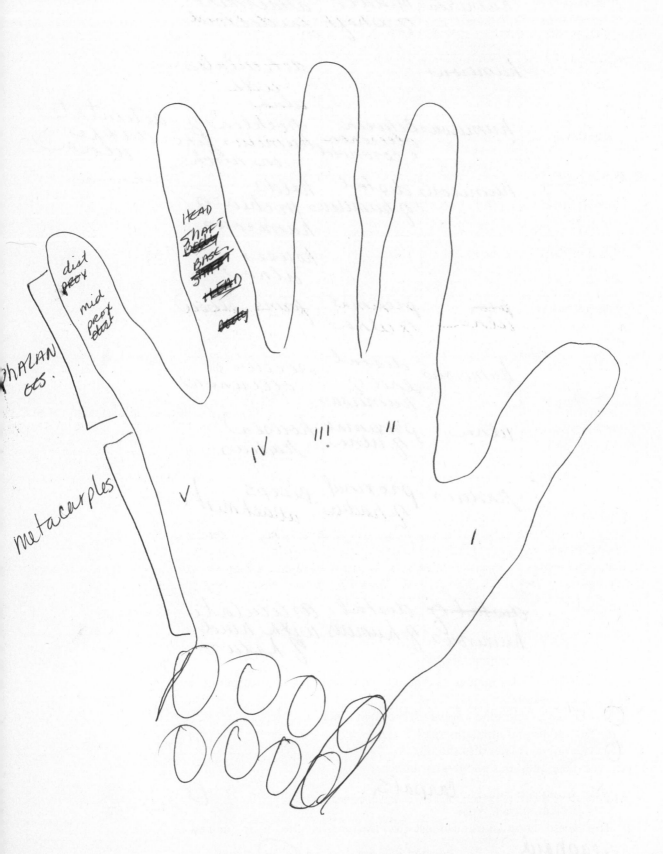

HEAD
SHAFT
~~Body~~
~~BASE~~
~~SHAFT~~
~~HEAD~~
~~Body~~

dist
PROX

mid
PROX
dist

PHALAN
ges.

metacarples

V

IV   III   II

I

6. Referring to the figure you just drew, complete this exercise about the hand. ☐4   **79**

   a. The bones composing the palm of the hand are called _META_ _CARPALES_ bones. There are _____ in each hand. The one on the thumb side is numbered (*I? V?*).

   b. Metacarpals articulate proximally with _DISTAL CARPALS_, laterally with _EACH OTHER_, and distally with _FLANGES_

   c. Finger bones are called _PHALANGES_. Each digit, except for the thumb, has _3_ bones; the thumb has _2_ phalanges.

C. Pelvic girdle

(pages 151–153)

1. Name the bones that form the pelvis.

   illum, pubis, ishium, Sacrum

   Which of these bones is/are part of the axial skeleton? ☐5

   SACRUM.

2. Contrast the two principal parts of the pelvis. Describe their location and name the structures which compose them.

   a. Greater (false) pelvis

   Superior of illia & Superior Sacrum. & wall of the abdomen.

   b. Lesser (true) pelvis

   inferior illia, Sacrum, coxyx & pubes

3. Look at the pelvic inlet on Figure 7-7 of your text and Figure LG 7-1. Find the brim of the true pelvis (*pelvic brim*) which demarcates the opening known as the pelvic inlet. This line is very (*smooth? irregular?*).

4. Now consider the *pelvic outlet*. With your fingertip, trace it on a skeleton, if available. It is very (*smooth? irregular?*). Why is the pelvic outlet so named? ☐6

5. Define *pelvimetry*. For what purpose is it used?

   measure of size of inlet & outlet of birth canal.

6. Answer the following questions about coxal bones. ☐7

   a. The coxal bones are also known as _____ or

   _____ .

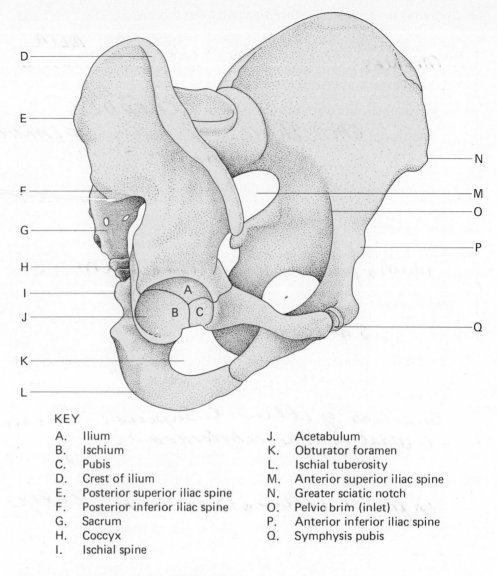

KEY

A. Ilium	J. Acetabulum
B. Ischium	K. Obturator foramen
C. Pubis	L. Ischial tuberosity
D. Crest of ilium	M. Anterior superior iliac spine
E. Posterior superior iliac spine	N. Greater sciatic notch
F. Posterior inferior iliac spine	O. Pelvic brim (inlet)
G. Sacrum	P. Anterior inferior iliac spine
H. Coccyx	Q. Symphysis pubis
I. Ischial spine	

Figure LG 7-1 Pelvic girdle. Color as directed.

b. Each coxal bone originates as three bones which fuse early in life. These three bones are the _____ , _____ , and _____ .

c. At what location do the three bones fuse?

d. The largest of the three bones is the _____ . A ridge along the superior border is called the iliac crest. Locate this on yourself.

e. The iliac crest ends anteriorly as the _____ _____ spine.

The crest ends posteriorly as the _____ _____ . This marking causes a dimpling of the skin just lateral to the sacrum, which can be used as a landmark for administering hip injections accurately.

7. Color Figure LG 7-1 according to the following code: *ilium,* yellow; *ischium,* red; *pubis,* blue; *sacrum,* green; *coccyx,* black.

8. Complete the table about markings of the coxal bones.

Marking	Location on Coxal Bone	Function
a. Greater sciatic notch		
b.		Supports most of body weight in sitting position
c. Auricular surface		
d.		Fibrocartilaginous joint between two coxal bones
e.		Socket for head of femur
f. Obturator foramen		

D. Lower extremities

(pages 153–163)

1. List the bones in the lower extremities from proximal to distal.

femur, knee cap, fibula, tibia tarsals, metatarsals, phalanges.

2. Locate each of the bones on your list on Figure 7-9, on yourself, and on a skeleton (if available).

3. Circle the correct term indicating locations of these parts of the lower extremity.

9

a. The head is the (*proximal? distal?*) epiphysis of the femur.

b. The greater trochanter is (*lateral? medial?*) to the lesser trochanter.

c. The intercondylar fossa is on the (*anterior? posterior?*) surface of the femur.

d. The tibial condyles are more (*concave? convex?*) than the femoral condyles.

e. The lateral condyle of the femur articulates with the (*fibula? lateral condyle of the tibia?*).

f. The tibial tuberosity is (*superior? inferior?*) to the patella.

g. The tibia is (*medial? lateral?*) to the fibula.

h. The outer portion of the ankle is the (*lateral? medial?*) malleolus which is part of the (*tibia? fibula?*).

4. The tarsal bone which is most superior in location (and which articulates with the tibia and fibula) is the _____ . The largest and strongest of the tarsals is the _____ .

5. Answer these questions about the arch of the foot.

a. How is the foot maintained in an arched position?

b. Describe the different arches of the foot.

c. What causes flatfoot?

6. Now that you have seen all of the bones of the appendicular skeleton, complete this table relating common and anatomical names of bones.

$\boxed{10}$

Common Name	Anatomical Name
a. Shoulder blade	
b.	Pollex
c. Collarbone	
d. Heel bone	
e.	Olecranon process
f. Kneecap	
g.	Tibial crest
h. Toes	
i. Palm of hand	
j. Wrist bones	

8

Articulations

In the past three chapters you have learned a great deal about the 206 bones in the body. Separated—or disarticulated—these bones would be no more than a disorganized pile. But bones are arranged in precise order and are held together in various ways so that they can function effectively. In this chapter you will look at different kinds of joints and the movements possible at each (Objectives 1–3). You will examine closely three selected joints: the shoulder, hip, and knee (4). You will also learn about a number of disorders involving joints (5–6), as well as applications to health and key medical terms (7).

Topics Summary

A. Classification
B. Fibrous joints, cartilaginous joints
C. Synovial joints: structure
D. Synovial joints: movements
E. Synovial joints: types
F. Selected articulations of the body
G. Applications to health; key medical terms

Objectives

1. Define an articulation and identify the factors that determine the degree of movement at a joint.

2. Contrast the structure, kind of movement, and location of fibrous, cartilaginous, and synovial joints.

3. Discuss and compare the movements possible at various synovial joints.

4. Describe selected joints of the body with respect to the bones that enter into their formation, structural classification, and anatomical components.

5. Define the causes and symptoms of common joint disorders, including arthritis, rheumatism, rheumatoid arthritis, osteoarthritis, gouty arthritis, septic arthritis, bursitis, and tendinitis.

6. Define dislocation and sprain.

7. Define key medical terms associated with joints.

88 **Learning Activities**

(page 166) **A.** Classification

1. Define the term *articulation* (*joint*).

1   2. Name three classes of joints based on function.

3. Name three classes of joints based on structure.

(pages 166–167) **B.** Fibrous joints, cartilaginous joints

1. Describe the general characteristics of *fibrous joints*.

2   2. Name three types of fibrous joints.

3. Complete the table about kinds of *sutures*.

Type of Suture	Description	Example
a.		Between parietal bones
b.	Margins of adjacent bones that overlap	
c. Plane suture		

4. Contrast sutures with these two other types of fibrous joints:
   a. Synostoses

   b. Syndesmoses

5. Compare these two types of cartilaginous joints with regard to structure: *synchondroses* and *symphyses*.

6. Complete the summary table about fibrous and cartilaginous joints.   3

Type of Joint	Structural Class	Functional Class	Example
a.	Fibrous		Between occipital and parietals
b.			Distal articulation of tibia and fibula
c. Gomphosis			
d.		Synarthrotic	Epiphyseal plate
e. Symphysis			

C. Synovial joints: structure   (pages 167–169)

1. What structural features of synovial joints make them more freely movable than fibrous or cartilaginous joints?   4

2. Complete the table about parts of a synovial joint.

Part of Joint	Structure	Function
a. Synovial cavity		
b.	Hyaline cartilage covering surfaces of articulating bones; does not bind bones together	
c. Fibrous capsule		
d.		Secretes synovial fluid
e. Synovial fluid		
f. Ligaments		

3. Contrast *fibrous capsule* with *ligaments*.

4. Label the parts of a synovial joint on Figure LG 8-1. Compare your labels to those in Figure 8-2 in the text.

5. What are *menisci?* Give an example of a joint that has menisci.

5

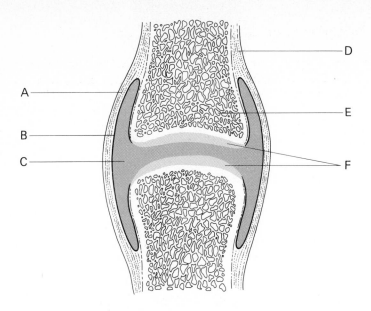

Figure LG 8-1 Structure of a synovial joint.

6. What are *bursae?* Where are they located? What is their function?

7. What is *bursitis?* What factors do you think might cause bursitis?

D. Synovial joints: movements   **(pages 169–174)**

1. The design of synovial joints permits free movement of bones. However, if bones moved too freely, they could move right out of their joint cavities (dislocation). Describe how each of these factors accounts for limitation of movement at synovial joints.
   a. Apposition of soft parts

b. Tension of ligaments

c. Muscle tension

2. Choose the description that fits the type of movement in each case. Not all answers will be used.

6

_____ a. Decrease in angle between anterior surfaces of bones (or between posterior surfaces at knee and toe joints)

_____ b. Simplest kind of movement that can occur at a joint; no angular or rotary motion involved; example: ribs moving against vertebrae

_____ c. State of entire body when it is in anatomical position

_____ d. Movement away from the midline of the body

_____ e. Movement of a bone around its own axis

_____ f. Position of foot when heel is on the floor and rest of foot is raised

Abd. Abduction
Add. Adduction
  C. Circumduction
  D. Dorsiflexion
  E. Extension
  F. Flexion
  G. Gliding
  I. Inversion
  P. Plantar flexion
  R. Rotation

3. Perform the action described. Then write in the name of the type of movement.

7

a. Describe a cone with your arm, as if you are winding up to pitch a ball; the movement is called _____.

b. Stand in anatomical position (palms forward). Now turn your palms backward. The action you just performed is called _____ _____.

c. Raise your shoulders, as if to shrug them. This movement is called _____ of the shoulders.

d. Stand on your toes. This action at the ankle joint is called _____.

e. Grasp a ball in your hand. Your fingers are performing the type of movement called _____.

4. Identify the kinds of movements shown in Figure LG 8-2. Write the name of the movement below each figure. Use the following terms: *abduction, adduction, extension, flexion,* and *hyperextension.*

8

1. After you have studied the types of joints (Exhibit 8-1 in your text), check your understanding by choosing the type of synovial joint that fits the description. (Answers may be used more than once.) |9|

    _____ a. Monaxial joint; only rotation possible

    _____ b. Joint between carpal and metacarpal of the thumb

    _____ c. Shoulder and hip joints

    _____ d. Spoollike surface articulated with concave surface

    _____ e. Monaxial joint; only flexion and extension possible

    _____ f. Biaxial joints (three answers)

B. Ball-and-socket
E. Ellipsoidal
G. Gliding
H. Hinge
P. Pivot
S. Saddle

2. All synovial joints are classified functionally as:
A. Diarthrotic
B. Synarthrotic
C. Amphiarthrotic

F. Selected articulations of the body (pages 175–198)

1. After you have studied the selected joints in Exhibits 8-2 through 8-11, do the following exercise. Match the joints in the list at right with the descriptions below. |10|

    _____ a. Joint between axial and appendicular skeleton

    _____ b. Joint between skull and vertebral column

    _____ c. Cartilaginous discs consisting of anulus fibrosus and nucleus pulposus cushion bones in these joints

    _____ d. Fibrocartilage disc separates joint cavity into superior and inferior compartments

    _____ e. Anterior and posterior longitudinal ligaments, as well as ligaments flava, add stability to these joints

    _____ f. Ankle joint

    _____ g. Cruciate ligament located here

    _____ h. Radial and ulnar collateral ligaments located here

AA. Atlantoaxial
AO. Atlantooccipital
IV. Intervertebral
LS. Lumbosacral
RC. Radiocarpal
TC. Talocrural
TM. Temporomandibular

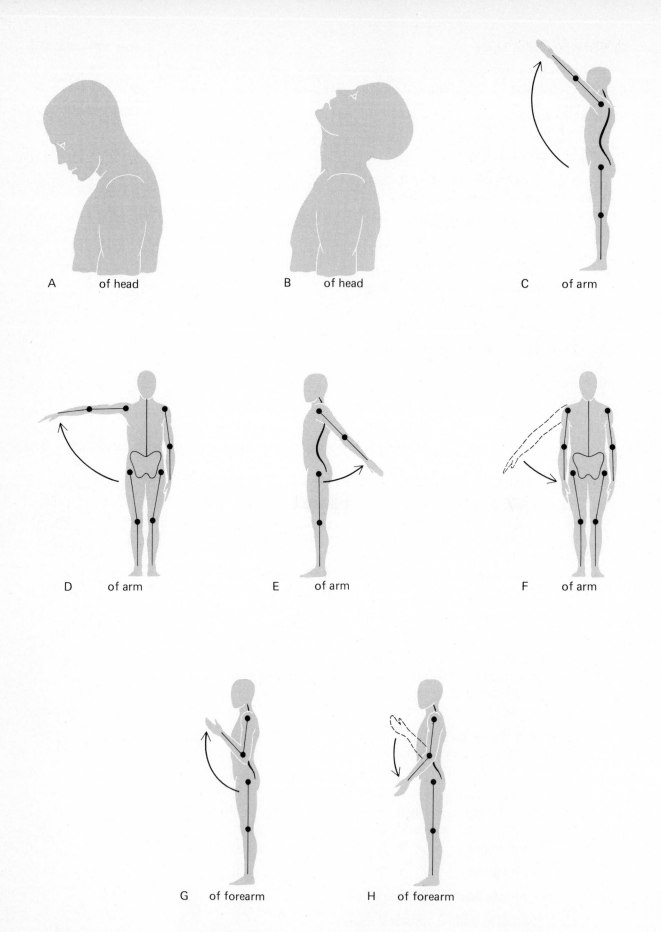

A   of head

B   of head

C   of arm

D   of arm

E   of arm

F   of arm

G   of forearm

H   of forearm

Figure LG 8-2 Movements at synovial joints. Letters refer to question $\boxed{8}$ and to questions in Chapter 10.

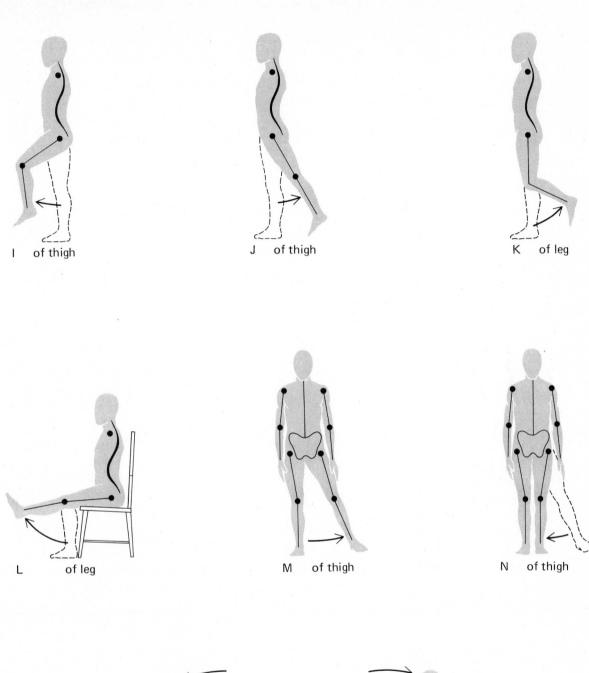

I   of thigh

J   of thigh

K   of leg

L   of leg

M   of thigh

N   of thigh

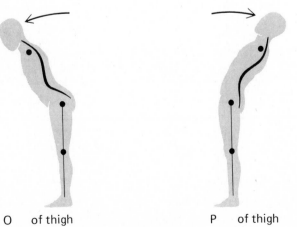

O   of thigh

P   of thigh

2. Discuss each of the joints listed below. Name the articulating bones. Include a description of the *articular capsule,* major *ligaments, bursae,* and other special structures associated with the joint.

   a. Shoulder

   b. Hip

   c. Knee

3. What is the *rotator cuff?* How is it involved with strength and stability of the shoulder joint?

11   4. What is the function of the *labrum?* In which joints is a labrum found?

5. Are the structures listed below associated with the shoulder, hip, or knee? $\boxed{12}$

 a. Medial and lateral menisci

 b. Anterior and posterior cruciate ligaments

 c. Iliofemoral ligaments

 d. Subdeltoid bursa

6. Further compare these three major synovial joints by completing the following exercise. $\boxed{13}$

 a. Which joint has the widest range of motion?

 b. Which one has most limited range of motion?

 c. Which is most stable?

 d. Which is least stable?

G. **Applications to health; key medical terms** (pages 199–203)

1. How are *arthritis* and *rheumatism* related? Choose the correct answer. $\boxed{14}$
 A. Arthritis is a form of rheumatism.
 B. Rheumatism is a form of arthritis.

2. List several forms of arthritis. Name one symptom common to all forms of this ailment.

3. Contrast *rheumatoid arthritis* with *osteoarthritis.* Write *More, Less, Yes,* or *No* for answers. $\boxed{15}$

	Rheumatoid Arthritis	Osteoarthritis
a. Which is more common?		
b. Which is more damaging?		
c. Is this an inflammatory condition?		
d. Is synovial membrane affected?		
e. Does articular cartilage degenerate?		
f. Does fibrous tissue join bone ends?		
g. Do bone spurs build up and limit joint movement?		

4. Briefly describe each of these disorders involving articulations.
   a. Gouty arthritis

   b. Septic arthritis

   c. Bursitis

   d. Tendinitis

   e. Dislocation

   f. Sprain

5. Write the meaning of each of the following word roots or suffixes. Then give one example of a term related to joints that contains the word root or suffix. The first one is done for you.
   a. *ectomy:* removal of; bursectomy

   b. *osis*

   c. *algia*

   d. *arthro*

   e. *chondro*

9

Muscle Tissue

In the past four chapters you have learned about the structural framework of the body. The skeleton provides support, and its jointed arrangement makes movement possible. However, movement can occur only when muscles pull on bones. In this chapter you will look closely at the structure and function of muscle tissue (Objectives 1–5), learn some principles of muscle contraction (6–8), and contrast skeletal muscle with smooth and cardiac muscle (9). You will also study disorders and key medical terms associated with the muscular system (10–12).

Topics Summary

A. Characteristics, functions, kinds of muscle tissue
B. Skeletal muscle tissue
C. Contraction
D. Cardiac muscle tissue, smooth muscle tissue
E. Applications to health and key medical terms

Objectives

1. List the characteristics and functions of muscle tissue.
2. Compare the location, microscopic appearance, nervous control, and functions of the three kinds of muscle tissue.
3. Define fascia, epimysium, perimysium, endomysium, tendons, and aponeuroses, and list their modes of attachment to muscles.
4. Describe the relationship of blood vessels and nerves to skeletal muscles.
5. Identify the histological characteristics of skeletal muscle tissue.
6. Describe the importance of the motor unit.
7. Define the all-or-none principle of muscular contraction.
8. Compare fast (white) muscle with slow (red) muscle.
9. Contrast cardiac muscle tissue with smooth muscle tissue.
10. Define such common muscular disorders as fibrosis, fibrositis, "charleyhorse," muscular dystrophy, and myasthenia gravis.
11. Compare spasms, cramps, convulsions, tetany, fibrillation, and tics as abnormal muscular contractions.
12. Define key medical terms associated with the muscular system.

(page 206) **A.** Characteristics, functions, kinds of muscle tissue

1. List four characteristics of muscle tissue. Write a sentence defining each.

2. What functions does muscle tissue perform?

3. Complete the table about kinds of muscle tissue.

Kind	Location	Striated or Smooth	Voluntary or Involuntary
a. Skeletal			
b.		Smooth	
c.	Only in the heart		

(pages 206–211) **B.** Skeletal muscle tissue

1. Contrast the kinds of fascia by indicating which of the following are characteristics of superficial fascia (S) or deep fascia (D).

1

_____ a. Immediately under the skin

_____ b. Composed of dense connective tissue

_____ c. Contains much fat, so serves as insulator of the body

_____ d. Also called subcutaneous layer

_____ e. Forms epimysium, perimysium, and endomysium that holds muscles into functional groups

2. Describe the extensive network of deep fascia that surrounds entire muscles, muscle bundles, and muscle cells and that connects muscles to bones or other muscles. Include in your description: *epimysium, perimysium, endomysium, tendon,* and *aponeurosis.*

3. What functions do tendon sheaths serve? Where are they located?

4. Explain why good nerve and blood supplies are required for normal muscle function.

5. Draw a diagram of a skeletal muscle fiber (cell). Label: *sarcolemma, sarcoplasm, nuclei, mitochondria, sarcoplasmic reticulum,* and *myofibrils.*

6. Carefully study a diagram of skeletal muscle tissue (Figure 9-3c in your text). Then check your understanding of the *sarcomere* and its components by doing this exercise.

2

a. In muscle tissue fibers are (*cells? intercellular material?*), whereas in connective tissue fibers are (*cells? intercellular material?*).

b. Each muscle fiber is surrounded by a membrane called the _____ and contains cytoplasm called _____ _____.

c. Nuclei and mitochondria within muscle cells are located close to _____.

d. What is *sarcoplasmic reticulum?*

e. T tubules run (*parallel? perpendicular?*) to sarcoplasmic reticulum.

f. A T tubule, along with sarcoplasmic reticulum on either side, is called a _____.

g. Each myofibril consists of bundles of thick and thin _____ _____ stacked in compartments called _____ _____.

h. The ends of sarcomeres are marked by (*A? H? I? Z?*) lines, which are the sites of (*T tubules? sarcoplasmic reticulum?*).

i. Each sarcomere is about 2.6 μm long. Of this, the central 1.6 μm is a dark band called the _____ band. The light colored bands at ends of the sarcomere, called _____ bands, total 1.0 μm.

j. Which part of the sarcomere contains only thick filaments? (*A band? I band? H zone?*) Which part contains only thin myofilaments? _____

k. (*Thick? Thin?*) myofilaments are composed of myosin.

l. Thin myofilaments are composed mostly of the molecule named _____, along with two other molecules named _____ and _____. All three of these molecules are (*carbohydrate? protein? lipid?*).

m. (*Actin? Myosin?*) molecules are shaped like rods with round heads called _____.

n. Each thick myofilament is surrounded by (*50? 12? 6? 2?*) thin myofilaments.

(pages 211–215) **C.** Contraction

1. Explain how muscle contraction occurs according to the sliding filament theory.

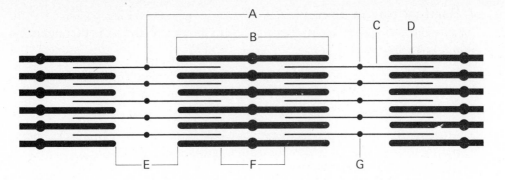

Figure LG 9-1 Detail of a sarcomere in relaxed muscle. Draw contracted muscle in space below.

2. Identify parts of the sarcomere in Figure LG 9-1 by matching each lettered structure in the figure to the correct term in the following exercise. 　3

　　_____ a. Sarcomere

　　_____ b. I band

　　_____ c. A band

　　_____ d. Thin myofilament

　　_____ e. Thick myofilament

　　_____ f. Z line

　　_____ g. H zone

In the space immediately below Figure LG 9-1, draw the sarcomere as it appears in the contracted state.

3. To describe the sarcomere during contraction, complete each statement with one of the following: lengthens, shortens, or stays the same length. 　4

　a. The sarcomere _____ .

　b. Each thick myofilament (A band) _____ .

　c. Each thin myofilament _____ .

　d. The I band _____ .

　e. The H zone _____ .

4. Name the parts of a *motor unit*.

5. Muscles that control precise movements have (*more? fewer?*) muscle fibers per motor unit than muscles controlling gross movements.

6. State the all-or-none principle.

This principle applies to (*individual motor units? an entire skeletal muscle such as the biceps?*).

5   7. List three factors that decrease the strength of a muscle contraction.

8. Define *threshold* (*liminal stimulus*).

9. Explain how muscle tone is produced, and state the significance of muscle tone in normal body function.

10. Define these terms and explain their clinical significance.
   a. Flaccid

   b. Atrophy

   c. Hypertrophy

11. Contrast fast and slow muscle types by completing this table.

Type of Muscle	Color	Presence of Myoglobin	Structural Differences	Example
a. Fast	White			
b. Slow		Yes		Gastroc-nemius

12. The substance in muscle that stores oxygen until oxygen is needed by mitochondria is _____ .   6

D. Cardiac muscle tissue, smooth muscle tissue   (pages 215–216)

1. Draw a diagram of a cardiac muscle fiber (cell). Then point out structural differences between this cell and the skeletal muscle fiber (cell) you drew in activity B5.

2. Draw a diagram of a single smooth muscle fiber (cell). How does it differ from fibers of skeletal and cardiac muscle?

3. Compare the two types of smooth muscle.

Muscle Type	Structure	Spread of Stimulus	Locations
a. Visceral		Impulse spreads and causes contraction of adjacent fibers	
b.			Blood vessels, iris of eye

(pages 218–220) **E. Applications to health; key medical terms**

1. Briefly describe each of these disorders.
   a. Fibrosis

b. Fibromyositis

c. Muscular dystrophy

d. Myasthenia gravis

2. Write the name of the disorder following the correct description. ⬜ 7

   a. Weakness of skeletal muscle due to ab-   Fibromyositis
     normality at neuromuscular junction: Fibrosis
                                        Fibrositis
    _____   Muscular dystrophy

   b. "Charleyhorse": _____   Myasthenia gravis
   c. Called lumbago if it occurs in lumbar re-

     gion: _____

   d. Formation of fibrous tissue where it does

     not belong: _____

   e. Inherited, muscle-destroying disease

     causing atrophy of muscle tissue: _____

_____

3. A number of muscular disorders can be identified by increased blood levels

of the enzyme called _____ .

4. Explain how *plasmapheresis* can help individuals with myasthenia gravis.

5. Define each of these types of abnormal muscle contractions.
   a. Spasm

   b. Cramp

   c. Convulsion

   d. Fibrillation

   e. Tic

6. Write the meaning of each of the following word roots:
   a. *myo*

   b. *pathos*

   c. *tonia*

   d. *tortus*

7. Define each of these medical terms related to muscles.
   a. Myology

   b. Myomalacia

10
The Muscular System

In this last chapter on the musculoskeletal system, you will learn how skeletal muscles, by means of their attachments to bones, produce the wide array of movements of the body (Objective 1). You will consider the significance of structural patterns of muscles and their arrangement in functional groups (2–4). You will study criteria which will facilitate your learning names of muscles (5). Then you will identify the major skeletal muscles of the body with regard to origin, insertion, action, and innervation (6–7). You will also study methods of drug administration, with emphasis on intramuscular injections (8).

Topics Summary

A. How skeletal muscles produce movement
B. Naming skeletal muscles
C. Principal skeletal muscles of facial expression; muscles that move the lower jaw, eyeballs, tongue, pharynx, larynx, and head
D. Principal skeletal muscles that act on the abdominal wall, muscles used in breathing, muscles of the pelvic floor and perineum
E. Principal skeletal muscles that move the shoulder girdle and upper extremity
F. Principal skeletal muscles that move the vertebral column
G. Principal skeletal muscles that move the lower extremity
H. Intramuscular injections

Objectives

1. Describe the relationship between bones and skeletal muscles in producing body movements.
2. Define a lever and fulcrum and compare the three classes of levers on the basis of placement of the fulcrum, effort, and resistance.
3. Discuss most body movements as activities of groups of muscles by explaining the roles of the prime mover, antagonist, and synergist.
4. Identify the various arrangements of muscle fibers in a skeletal muscle and relate the arrangements to the strength of contractions and range of movements.
5. Define the criteria employed in naming skeletal muscles.
6. Identify the principal skeletal muscles in different regions of the body by name, origin, insertion, action, and innervation.
7. Identify surface anatomy features of selected skeletal muscles.
8. Compare the common sites of intramuscular injection.

(pages 223–225) **A. How skeletal muscles produce movement**

1. What structures constitute the *muscular system?*

1

2. Flex your left forearm. Now carefully examine that movement by completing this exercise.

   a. Your forearm serves as a _____ since it is a rigid rod that moves around a fixed point. The fixed point, known as the

   _____ , is your elbow joint.

   b. If you were holding a weight in your left hand as you flexed your forearm, the weight plus your forearm would serve as the (*effort? fulcrum? resistance?*) during this movement.

   c. The effort to move your forearm is provided by _____

   _____ .

   d. Each skeletal muscle is attached to at least two bones. As the muscle shortens, one bone stays in place, and so is called the

   _____ end of the muscle. What bone do you think might serve as the origin during flexion of the arm? _____

   _____

   e. The attachment of the muscle to the bone that moves is called the

   _____ end of the muscle. In this case the insertion is more likely to be the (*humerus? radius or ulna?*).

   f. Consider the locations of fulcrum, effort, and resistance when your forearm is flexed. Which is most proximal? _____ Which

   most distal? _____ This is an example of a (*first-? second-? third-?*) class lever?

   g. Which of these is also a third-class lever? (*Seesaw? Wheelbarrow? Shovel?*) Third-class levers are the (*most? least?*) common class of levers in the body.

   h. The belly of the muscle causing flexion of the forearm will most likely cover the (*humerus? radius or ulna?*).

   i. The muscle that contracts to cause this action is called a

   _____ . An example of a prime mover in this case

   would be the _____ muscle.

   j. The triceps brachii must relax as the biceps brachii flexes the forearm. The triceps is an extensor. Since its action is opposite to that of the biceps, the triceps is called (*a synergist? an agonist? an antagonist?*) of the biceps.

   k. What would happen if the flexors of your forearm were functional, but not the antagonistic extensors?

l. What function does a synergistic muscle perform?

3. Describe the four basic patterns of muscle fiber arrangement within fasciculi (bundles). Give an example of each.
   a. Parallel

   b. Convergent

   c. Pennate

   d. Circular

4. Correlate fascicular arrangement with muscle power and range of motion of muscles.

   [2]

   a. A muscle with (*many? long?*) fibers will tend to have great strength. An example is the (*parallel? pennate?*) arrangement.
   b. A muscle with (*many? long?*) fibers will tend to have great range of motion. An example is the (*parallel? pennate?*) arrangement.

B. Naming skeletal muscles

(page 226)

1. If you look closely at the name of a muscle, you will find that most muscle names provide a good description of the muscle. For each of the following, indicate the type of clue that each part of the name gives.

   [3]

   _____ a. Rectus abdominis   A. Action
   _____ b. Gluteus maximus   D. Direction of fibers
                           L. Location
   _____ c. Biceps brachii   N. Number of heads or origins
   _____ d. Sternocleidomastoid   P. Points of attachment of origin and insertion
   _____ e. Adductor longus   S. Size or shape

2. Read the following descriptions of muscles, noting italicized words that give direct clues to muscle names. Guess the names of these muscles. As you proceed with your study of specific muscles, identify muscles whose names match these descriptions. Then return to this exercise and see if the names you gave initially were correct.

   a. A *circular* muscle around the *mouth*

   b. *Elevates* the *upper lip*

   c. Attached to the *styloid* process of the temporal bone and also to the *tongue*

   d. A muscle whose fibers run vertically (*straight*) up and down the *abdomen*

   e. A muscle that *lifts* the *scapula*

   f. The *broadest* muscle of the *back*

   g. A *triangular* muscle

   h. Located *above* the *spine* of the scapula

   i. A *round* muscle that *pronates* the forearm

   j. *Flexes* the *carpals* and lies along the *ulna*

   k. *Tenses* a *wide* band of *fascia* on the thigh

   l. A *four-headed* muscle lying over the *femur*

   m. A *large* muscle lying on the *lateral* side of the thigh

   n. A *slender* muscle

   o. A muscle that permits you to sit cross-legged, the position of *tailors*

   p. A muscle that lies *anterior* to the *tibia*

   q. A *long* muscle that *extends* the toes (*digits*)

(pages 226–242) **C. Principal skeletal muscles of facial expression; muscles that move the lower jaw, eyeballs, tongue, pharynx, larynx, and head**

   1. Name the muscles that answer each of the following descriptions.
      a. Elevates upper eyelid

      b. Elevates upper lip

      c. Produces a frowning expression

      d. A flat muscle that causes pouting action; draws lip downward and backward

e. Major cheek muscle; allows you to blow air out of mouth and produce sucking action

f. Allows you to show surprise by raising your eyebrows and forming horizontal forehead wrinkles

g. Muscle surrounding opening of your mouth; allows you to use your lips in kissing and in speech

2. Essentially all of the muscles controlling facial expression receive nerve impulses via the _____ nerve, which is cranial nerve (*III? V? VII?*).

3. Most of the muscles involved in chewing are ones that help you to (*open? close?*) your mouth. Think about this the next time you go to the dentist and try to hold your mouth open for a long time, with only your

_____ muscles to force your mouth wide open. | 4 |

4. Name the two large muscles that help you close your mouth forcefully, as in chewing. Note that both of them act by (*lowering the maxilla? elevating the mandible?*).

a. _____ covers your temple.

b. _____ covers the ramus of the mandible.

5. With the help of Figures 10-4 and 10-5 in your text and a mirror, locate each of the major facial muscles.

6. Complete this exercise about muscles that move the eyeballs. | 5 |

a. Why are these muscles called *extrinsic* eyeball muscles?

b. All of the *rectus* muscles move the eyeball in the direction (*opposite to? the same as?*) that given in the muscle name. For example, the superior rectus moves the eyeball (*superiorly? inferiorly?*).

c. The superior and inferior oblique muscles both move the eyeballs (*medially? laterally?*) as well as toward the direction (*opposite to? same as?*) that given in the name. For example, the inferior oblique moves the eye-

ball _____ .

d. Only one muscle moves the eyeball at all in the medial direction. This is

the _____ muscle.

7. Identify locations of three bony structures relative to your tongue. Tell whether they are superior or inferior and anterior or posterior to your tongue.

a. Styloid process of temporal bone

b. Hyoid bone

c. Chin (anterior portion)

8. Remembering the locations of these bony points and that muscles move a structure (tongue) by pulling on it, determine the direction that the tongue is pulled by each of these muscles.

☐6

   a. Styloglossus

   b. Hyoglossus

   c. Genioglossus

9. Name the muscles that cause the following actions.
   a. Produces tension on vocal folds of larynx during speech

   b. Elevates thyroid cartilage during swallowing to help prevent passage of food into larynx

   c. Constricts pharynx

10. Using a mirror, find the attachments of your left sternocleidomastoid muscle. The muscle contracts when you pull your chin down and to the right; this diagonal muscle of your neck will then be readily located. Note that the left sternocleidomastoid pulls your face towards the (*same? opposite?*) side. It also (*flexes? extends?*) the head.

11. On which surface of the neck would you expect to find extensors of the head? (*Anterior? Posterior?*) Name three of these muscles. Note that these muscles are the most superior muscles of the columns of extensors of the vertebrae. Find them on Figure 10-18 in your text.

☐7

12. Refer to Figure LG 8-2. Write in the names of muscles that produce the actions *A* and *B*.

(pages 243–251) **D. Principal skeletal muscles that act on the abdominal wall, muscles used in breathing, muscles of the pelvic floor and perineum**

1. Each half of the abdominal wall is composed of (*two? three? four?*) muscles. Describe these in the exercise below.

   a. Just lateral to the midline is the rectus abdominis muscle. Its fibers are

   (*vertical? horizontal?*), attached superiorly to the _____

   and inferiorly to the _____ . Contraction of this muscle permits you to bend (*forward? backward?*).

   b. Name the remaining abdominal muscles that form the sides of the abdominal wall. List them from most superficial to deepest.

c. Do all three of these muscles have fibers running in the same direction? _____ Of what advantage is this?

8

2. List three sets of muscles used during breathing.

3. Answer these questions about muscles used for breathing.

9

   a. The diaphragm is _____-shaped. Its oval origin is located

   _____.
   Its insertion is not into bone, but rather into dense connective tissue forming the roof of the diaphragm; this tissue is called the

   _____ .

   b. Contraction of the diaphragm flattens the dome, causing the size of the thorax to (*increase? decrease?*), as occurs during (*inspiration? expiration?*).

   c. The name *intercostals* indicates that these muscles are located

   _____ . Which set are used during expiration? (*Internal? External?*)

4. The pelvic floor consists of muscles attached to the bony pelvic outlet. Name these muscles and state functions of the pelvic floor.

5. Draw a diamond about the size and shape of the *perineum*. (Refer to page 650 in your text.)

  a. Label the four points that demarcate the perineum.
  b. Label the *urogenital triangle* and the *anal triangle*.
  c. What openings are located in these triangles in males? In females? Draw and label these in contrasting colors on your diagram.
  d. Draw in and label the following muscles as they appear in females: *bulbocavernosus, ischiocavernosus, levator ani,* and *sphincter ani externus.* Note which of these form parts of the penis, and so will be considerably larger in males than in females.

6. Define the *pelvic diaphragm.*

(pages 252–267) **E. Principal skeletal muscles that move the shoulder girdle and upper extremity**

1. On Figures LG 10-1 and LG 10-2, draw in pencil and label the muscles listed in the table below. Draw superficial muscles on the left side and deep muscles on the right side of the figures. Be sure to locate precise origins and insertions. Mark *O* and *I* on the origin and insertion ends of each muscle. Think about how origins and insertions determine what movements occur. Then use the spaces provided here for writing principal actions of each muscle.

Muscle	Action
a. Subclavius	
b. Pectoralis minor	
c. Serratus anterior	
d. Trapezius	
e. Levator scapulae	
f. Rhomboideus major	
g. Rhomboideus minor	

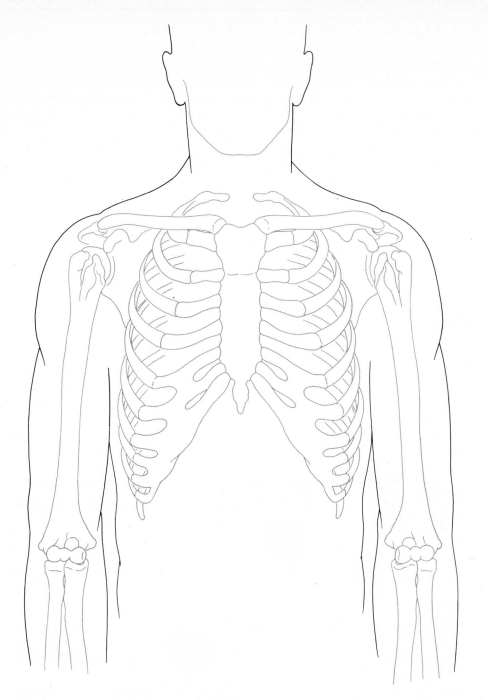

Figure LG 10-1 Anterior view of part of the skeleton. Draw and label muscles as directed.

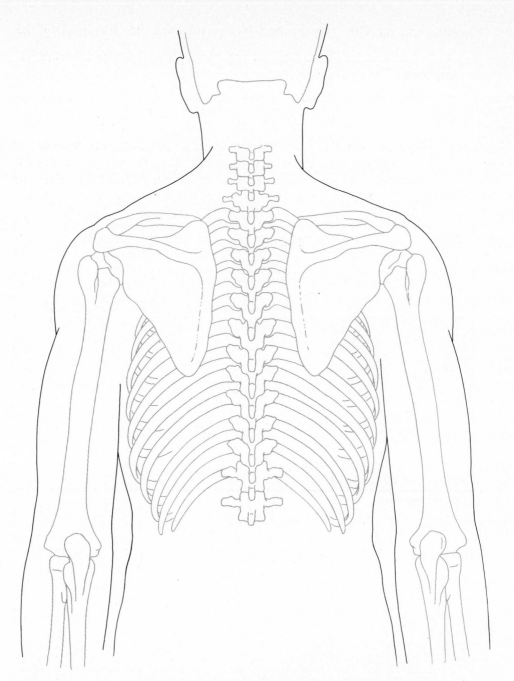

Figure LG 10-2 Posterior view of part of the skeleton. Draw and label muscles as directed.

2. All of the muscles listed in the table are involved with movement of the

_____ . All are (*superficial? deep?*) except the trapezius and parts of serratus anterior.

3. Using a red pencil or pen (for contrast) add, to these figures, drawings of the muscles listed below. Again, draw superficial muscles on the left and deep muscles on the right. Label and mark *O* and *I*. List actions in the table.

Muscle	Action
a. Pectoralis major	
b. Latissimus dorsi	
c. Deltoid	
d. Supraspinatus	
e. Infraspinatus	
f. Teres major	
g. Teres minor	

4. All of the muscles listed in the above table are directly involved with movement of the (*shoulder girdle? humerus? radius/ulna?*). All of these muscles are (*superficial? deep?*) with the exception of the supraspinatus and infraspinatus.
5. Using green pencil or pen, draw and label muscles that move the forearm; use the right sides of Figures LG 10-1 and LG 10-2. Follow the same directions as above for marking *O* and *I*. Then list actions here.

6. Try to locate each of the muscles named in the above three lists:
   a. On a skeleton, find origins and insertions and visualize actions.
   b. On yourself or on a partner, observe surface anatomy and actions.

7. Write the name of one or two muscles that produce each of the actions *C-H* shown in Figure LG 8-2. Write names next to each figure.

|10|

|11| 8. Complete this exercise about muscles that move the wrist and fingers.

   a. Examine your own forearm, palm, and fingers. There is more muscle mass on the (*anterior? posterior?*) surface. You therefore have more muscles that can (*flex? extend?*) your wrist and fingers.

   b. Locate the *flexor carpi ulnaris* muscle on Figure 10-17 in your text. What

   action does it have other than flexion of the wrist? _____

   _____ What muscles would you expect to abduct the wrist?

   c. What is the difference in location between *flexor digitorum superficialis* and *flexor digitorum profundus?*

   d. What muscle helps you to point (extend) your index finger?

|12| 9. Write the names of one or more muscles that fit these descriptions.

   a. Covers most of the posterior of the humerus

   b. Turns your hand from palm down to palm up position

   c. Originates from upper eight or nine ribs; inserts on scapula; moves scapula laterally

   d. Two muscles other than the biceps brachii that flex the forearm

   e. Used when a baseball is grasped

f. Antagonist to levator scapulae

g. Largest muscle of the chest region; used to throw a ball in the air (flex humerus) and to adduct arm

h. Raises or lowers scapulae, depending on which portion of the muscle contracts

i. Controls action at the elbow for a movement such as the downstroke in hammering a nail

j. Hyperextends the humerus, as in doing the "crawl" stroke in swimming

F. **Principal skeletal muscles that move the vertebral column**   **(pages 268–269)**

1. Describe the muscles that comprise the sacrospinalis group.

   a. Another name for the sacrospinalis muscle is the _____

      _____ .

   b. The muscle consists of three groupings: _____ (lateral), _____ (intermediate), and _____

      _____ (medial).

   c. In general, these muscles have attachments between: _____

      _____ .

   d. They are (*flexors? extensors?*) of the vertebral column, and so are (*synergists? antagonists?*) of the rectus abdominis muscles.

2. Explain why it is common for women in their final weeks of pregnancy to experience frequent back pains.   13

G. Principal skeletal muscles that move the lower extremity

1. Complete the table about muscles that move the thigh. Locate origins and insertions on a skeleton (if available), and try to determine locations of these muscles on yourself.

Muscle	Origin	Insertion	Actions
a. Psoas major			
b. Iliacus			
c. Gluteus maximus			
d. Gluteus medius			
e. Gluteus minimus			
f. Tensor fasciae latae			
g. Adductors			

2. Two large groups of muscles are involved with movements at the knee joint. Describe them in this exercise.

   a. Quadriceps femoris is the main muscle mass on the (*anterior? posterior?*) surface of the thigh.

   b. The name of this muscle mass indicates it has _____ heads of origin. All four converge to insert on the _____ bone by means of the patellar ligament. All four therefore cause (*flexion? extension?*) of the leg.

c. Three of the heads of the quadriceps originate on the _____ (bone). Their names are _____ , _____ , and _____ _____ . Since they cross only one joint, their only action is extension of the (*thigh? leg?*).

d. The fourth head, named the _____ , originates on the _____ spine of the _____ bone. Since it crosses the hip joint, it has the additional action of _____ the thigh.

e. The hamstrings are (*synergists? antagonists?*) of the quadriceps. Hamstrings are located on the (*anterior? posterior?*) surface of the thigh.

f. Why are the hamstrings so named?

g. The three muscles in the hamstring group are _____ _____ , _____ , and _____ .

h. All three of these muscles originate on the _____ bone and insert just below the knee. The actions they cause are the (*flexion? extension?*) of the thigh and the (*flexion? extension?*) of the knee.

3. Look at the muscles of the right hip and thigh in Figure 10-19. Now examine Figure LG 10-3.
   a. Identify locations of the major muscle groups of the right thigh in this cross section. Color the groups: adductors, blue; quadriceps femoris, red; hamstrings, black.
   b. Identify each of the muscles in Figure LG 10-3.

4. Sit in the "tailor's position": place the outer ankle of your right leg on top of your left knee. The muscle that crosses the thigh obliquely and causes this action of the right leg is _____ . $\boxed{14}$

5. Identify muscles that cause each action *I-P* shown in Figure LG 8-2. Write the muscle names next to each diagram. $\boxed{15}$

6. Perform these actions of your foot. Feel which muscles are contracting. Name two or more muscles that cause each action.
   a. Stand on your toes.

   b. Stand on your heels with toes turned up.

   c. Evert your foot.

   d. Invert your foot.

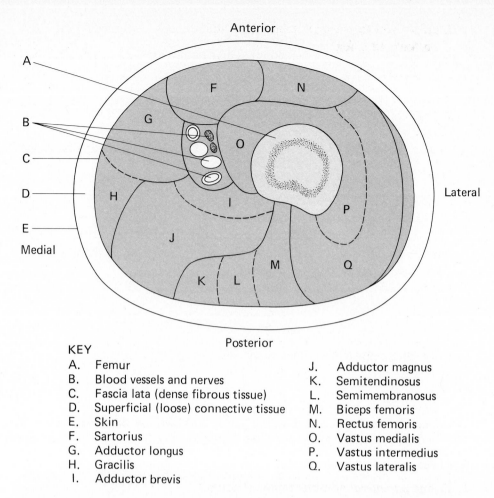

Anterior

Lateral

Medial

Posterior

KEY

A. Femur
B. Blood vessels and nerves
C. Fascia lata (dense fibrous tissue)
D. Superficial (loose) connective tissue
E. Skin
F. Sartorius
G. Adductor longus
H. Gracilis
I. Adductor brevis

J. Adductor magnus
K. Semitendinosus
L. Semimembranosus
M. Biceps femoris
N. Rectus femoris
O. Vastus medialis
P. Vastus intermedius
Q. Vastus lateralis

Figure LG 10-3 Cross section of right thigh midway between hip and knee joints. Structures A to E are nonmuscle, F to Q are skeletal muscles of the thigh. Color as directed.

7. Choose the names of the muscles that fit the descriptions. Write one muscle name on each line provided.

16

a. Allow you to touch your toes by causing flexion of your thigh:

_____ ,

_____ ,

_____

b. Allow you to kick your leg backwards (hyperextension of thigh):

_____ ,

_____ ,

_____

c. Become large in dancers because they stand on their toes often:

_____ ,

_____

Biceps femoris
Gastrocnemius
Gluteus maximus
Gluteus medius
Iliacus
Obturator internus
Psoas major
Rectus femoris
Semitendinosus
Soleus
Tensor fasciae latae
Vastus lateralis

d. Allow you to place your heel on your buttocks (flex leg):

_____,

_____

e. Laterally rotate thigh, an important action in normal walking:

_____,

_____

f. Antagonists of the hamstrings (with regard to action at knee joint):

_____,

_____

g. Form the "calf" of the leg:

_____,

_____

h. Form much of the mass of the buttocks:

_____,

_____

H. Intramuscular injections

(pages 281–282)

1. Define _parenteral_ administration of drugs.

2. List and define five parenteral routes of administration of drugs.

3. List reasons why intramuscular injections may be preferable to other parenteral methods.

17 4. Why is the gluteus medius a safer site for intramuscular injection than the gluteus maximus?

5. Name two other muscles commonly used for intramuscular injections.

Answers to Numbered Questions in Learning Activities

1 (a) Lever, fulcrum. (b) Resistance. (c) Muscles. (d) Origin; a bone proximal to the fulcrum: humerus or scapula. (e) Insertion, radius or ulna. (f) Fulcrum, resistance, third. (g) Shovel, most. (h) Humerus. (i) Prime mover or agonist, biceps brachii or brachialis. (j) An antagonist. (k) Your forearm would stay fixed in the flexed position. (l) It helps fix or stabilize the origin bone.

2 (a) Many, pennate. (b) Long, parallel.

3 (a) D, L. (b) L, S. (c) N, L. (d) P. (e) A, S.

4 Close, lateral pterygoid.

5 (a) They are outside of the eyeballs, not intrinsic like the iris. (b) The same as, superiorly. (c) Laterally, opposite to, superiolateral. (d) Medial rectus.

6 (a) Up and back. (b) Down. (c) Down and forward.

7 Sternocleidomastoid (*A*), three capitis muscles (*B*).

8 No. Strength is provided by the three different directions of the three muscles.

9 (a) Dome, around the bottom of the rib cage and on lumbar vertebrae, central tendon. (b) Increase, inspiration. (c) Between ribs (costa), internal.

10 Pectoralis major and coracobrachialis, also anterior portion of deltoid (*C*); deltoid and supraspinatus (*D*); latissimus dorsi and teres major (*E*); pectoralis major, latissimus dorsi, and teres major (*F*); biceps brachii, brachialis, and brachioradialis (*G*); triceps brachii (*H*).

11 (a) Anterior, flex. (b) Adducts wrist, flexor and extensor carpi radialis. (c) Superficialis is more superficial, and profundus lies deep. (d) Extensor indicis.

12 (a) Triceps brachii. (b) Supinator and biceps brachii. (c) Serratus anterior. (d) Brachialis and brachioradialis. (e) Flexor digitorum superficialis and profundus. (f) Pectoralis minor and lower fibers of trapezius. (g) Pectoralis major. (h) Trapezius. (i) Triceps brachii. (j) Latissimus dorsi.

11

The Cardiovascular System: Blood

In the next few chapters you will learn about a number of systems which help the human body to maintain homeostasis on a day-to-day basis. In Chapter 11 you will begin the study of the cardiovascular system with an examination of the blood. You will contrast roles of blood with those of lymph and interstitial fluid (Objectives 1, 11). You will identify components of blood and describe their origins, structural features, functions, and counts (2–10). You will study some important blood disorders and the medical terminology associated with blood (12–15).

Topics Summary

A. Types of body fluids; physical characteristics and functions of blood

B. Formed elements: types and origin

C. Formed elements: erythrocytes

D. Formed elements: leucocytes

E. Formed elements: thrombocytes

F. Plasma

G. Interstitial fluid and lymph

H. Applications to health; key medical terms

Objectives

1. Define the principal physical characteristics of blood and its functions in the body.

2. Compare the origins of the formed elements in blood and the reticuloendothelial cells.

3. Discuss the structure of erythrocytes and their function.

4. Define erythropoiesis and describe erythrocyte production and destruction.

5. Explain the importance of a reticulocyte count in the diagnosis of abnormal rates of erythrocyte production.

6. List the structural features and types of leucocytes.

7. Explain the significance of a differential count.

8. Discuss the role of leucocytes.

9. Discuss the structure and importance of thrombocytes.

10. List the components of plasma and explain their importance.

11. Compare the location, composition, and functions of interstitial fluid and lymph.

12. Contrast the causes and clinical symptoms of hemorrhagic, hemolytic, aplastic, and sickle cell anemia.

13. Define polycythemia and describe the importance of hematocrit in its diagnosis.

14. Identify the clinical symptoms of infectious mononucleosis and leukemia.

15. Define key medical terms associated with blood.

Learning Activities

(page 286) **A. Types of body fluids; physical characteristics and functions of blood**

1. Explain why specialized cells such as muscle or gland cells are dependent on the fluids of their *internal environment* for the maintenance of homeostasis.

2. Describe relationships among these three body fluids: blood, interstitial fluid, and lymph. (It may help to refer to Figure 11-3 in your text and Figure LG 11-1.)

3. Name the structures that comprise the:
   a. Cardiovascular system

   b. Lymphatic system

4. Describe these characteristics of blood.
   a. Viscosity

   b. pH

   c. NaCl concentration

5. Answer these questions about blood volume.　　　　　　　　　　|1|

   a. Blood constitutes about _____ percent of body weight. Mr. Jacob is a healthy 150-lb man. His body contains about _____ lb of blood.

   b. Mr. Jacob's body contains about _____ liters of blood. One liter equals about _____ pints. So, when Mr. Jacob donates a pint of blood, he is giving about _____ percent of his blood.

6. Name six substances carried by blood.

7. List five functions of blood other than transport.

B. **Formed elements: types and origin**　　　　　　　　　**(pages 286–288)**

1. Blood is a typical connective tissue in that it has (*much? little?*) intercellular material and relatively (*many? few?*) cells. The intercellular material of blood is a liquid named _____ . List the types of cells and other formed elements in blood.

2. The process of blood formation is called _____ .
3. Name the locations of blood formation in the fetus.

4. Contrast *myeloid tissue* and *lymphoid tissue*.

5. Explain why *hemocytoblasts* might be called "forefathers" of all blood cells.

2 6. Dead blood cells are removed by _____ cells. In which organs are large numbers of these "clean-up cells" located?

(pages 288–289) **C. Formed elements: erythrocytes**

1. Erythrocytes are also called (*red? white?*) blood cells.
2. Draw a mature red blood cell. Explain the functional advantage offered by the shape of a red blood cell.

3. What accounts for the red color of erythrocytes?

4. Describe how and where erythrocytes carry:
   a. Oxygen

   b. Carbon dioxide

5. The average life of a red blood cell is about _____ days. Briefly explain the reasons for the death of erythrocytes.

6. Circle the locations where red blood cells are normally produced in the adult.

Epiphyses of all bones   Bodies of vertebrae
Proximal epiphyses of both   Spleen
   femur and humerus bones   Diploe of cranial bones
Liver   Ribs
Sternum

7. Describe the stages of *erythropoiesis*. Indicate at what point the nucleus is lost.

8. Explain the roles of the following in erythropoiesis.
   a. Oxygen deficiency (hypoxia)

   b. Erythropoietin

   c. Iron

   d. Amino acids

   e. Vitamin B_{12}

   f. Intrinsic factor

D. Formed elements: leucocytes

4

1. State three structural characteristics of leucocytes which distinguish them from erythrocytes.

2. State two general functions of leucocytes.

3. Contrast the five types of white blood cells (WBCs) by completing the table.

Type of WBC	Percent of Total WBCs	Granular or Agranular	Source	Functions	Diagram of Cell Type
a.	60–70				
b.				Become plasma cells which produce antibodies	
c.	3–8		Lymphoid tissue		
d. Eosinophils					
				Become mast cells	

5

4. Complete this exercise about *antigens* and *antibodies*.

a. Antigens are defined as substances that _____ . Most are composed of (*carbohydrate? lipid? protein?*) Most (*are? are not?*) synthesized by the body. Several examples of antigens are

_____ .

b. Antigens stimulate _____ cells to form plasma cells which produce _____ . Chemically, these are all globulin-type _____ . They function to (*activate? inactivate?*) antigens.

5. State several examples of *antigen-antibody responses.*

6. Draw a box 1 mm on a side. Answer these questions about the number of cells in a cubic space that size within the human bloodstream.

| 6 |

a. There are normally about _____ red blood cells in the blood within a space this size.

b. There are normally about _____ white blood cells per cubic millimeter.

c. The ratio of red blood cells to white blood cells is normally about _____ to 1.

7. What accounts for the short life span of white blood cells?

E. Formed elements: thrombocytes   (pages 290–291)

1. Give the following information about thrombocytes.
   a. State another name for thrombocytes.

   b. Tell how they form.

   c. Describe their structure and size.

   d. What is their average life span?

   e. How many are usually present in 1 mm^3?

2. What is the major function of platelets?

(page 291) **F. Plasma**

1. Define *plasma.*

2. Match the names of each of these components of plasma with their descriptions.

| 7 |

_____ a. Makes up about 92 percent of plasma

_____ b. Regulatory substances carried in blood

_____ c. Ions such as Na$^+$, K$^+$, and HCO$_3^-$ carried in plasma

_____ d. A plasma protein; provides blood with viscosity so it maintains osmotic pressure

_____ e. A protein used in clotting

_____ f. Antibody protein

_____ g. Respiratory gases

_____ h. Food substances carried in blood

A. Albumin
E. Electrolytes
F. Fibrinogen
G. Globulin
GAF. Glucose, amino acids, and fats
HE. Hormones and enzymes
OC. Oxygen and carbon dioxide
W. Water

(pages 291–292) **G. Interstitial fluid and lymph**

| 8 |

1. Describe the types of fluid in the body by completing these statements and Figure LG 11-1.

a. Fluid inside cells is known as _____ fluid. Color areas containing this type of fluid yellow.

b. The fluid located in spaces between cells is called _____

_____. It surrounds and bathes cells and is one form of (*intracellular? extracellular?*) fluid. Color the spaces containing this fluid light green.

c. Another form of extracellular fluid is that located in _____

_____ vessels and _____ vessels. Color these areas dark green.

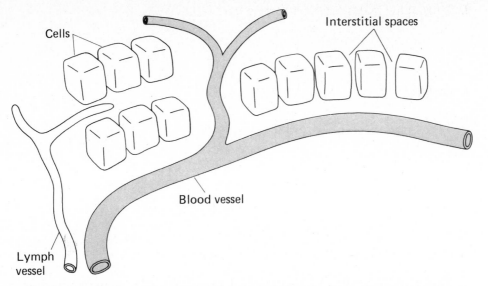

Figure LG 11-1 Internal environment of the body: types of fluid. Complete as directed.

2. Contrast these two body fluids in the table.

	Interstitial Fluid and Lymph	Plasma
a. Amount of protein		
b. Presence of blood cells		
c. Location		

H. Applications to health; key medical terms   (pages 292–295)

1. Define *anemia*. Contrast *pernicious anemia* with *nutritional anemia*.

2. List causes for each of these types of anemia.
   a. Hemorrhagic

   b. Hemolytic

c. Aplastic

d. Sickle cell

3. Define *hematocrit*.

Normal hematocrit value for males is _____ , for females _____ . Persons with polycythemia have (*increased? decreased?*) hematocrit.

4. Leukemia is a form of cancer involving abnormally high production of (*mature? immature?*) blood cells.

5. Write the meaning of each word root or suffix. Then give an example of a word related to blood that contains the word root or suffix.
   a. *Hemo*

   b. *Emia*

   c. *Thrombo*

   d. *Rrhage*

6. Contrast terms in each pair.
   a. Thrombus—embolus

   b. Direct transfusion—indirect transfusion

   c. Hemolysis—plasmolysis

12

The Cardiovascular System: The Heart and Blood Vessels

Now that you have examined the main transport fluid of the body—the blood—you will follow the pathway that it takes. First you will look closely at the structure that exerts a pumping force on the blood—the heart. You will study the normal heart (Objectives 1–8) as well as abnormalities of the heart (18–23). Then you will consider the closed system of tubules that serve as transport vessels or as exchange sites between blood and all body tissues (9–17). Finally you will learn about disorders related to blood vessels (24–25) and key medical terms (26).

Topics Summary

A. Location, size, and shape
B. Pericardium, walls and chambers, great vessels, valves
C. Conduction system, cardiac cycle, sounds, control
D. Blood vessels
E. Circulatory routes: systemic arteries and veins
F. Circulatory routes: coronary, hepatic portal, pulmonary, and fetal circulation
G. Applications to health; key medical terms

Objectives

1. Identify the position of the heart in the mediastinum.
2. Distinguish between the structure and location of the fibrous and serous pericardium.
3. Contrast the structure of the epicardium, myocardium, and endocardium.
4. Identify the blood vessels, chambers, and valves of the heart.
5. Explain the initiation and conduction of nerve impulses through the electrical conduction system of the heart.
6. Contrast the sounds of the heart and their clinical significance.
7. Discuss the surface anatomy features of the heart.
8. Contrast the effects of sympathetic and parasympathetic stimulation of the heart.
9. Contrast the structure and function of arteries, arterioles, capillaries, venules, and veins.
10. Identify the principal arteries and veins of systemic circulation.
11. Discuss the route of blood in coronary circulation.
12. Compare angina pectoris and myocardial infarction as abnormalities of coronary circulation.
13. Trace the route of blood involved in hepatic portal circulation and explain its importance.
14. Identify the major blood vessels of pulmonary circulation.
15. Contrast fetal and adult circulation.
16. Describe the importance of cerebral circulation.
17. Explain the fate of fetal circulation structures once postnatal circulation is established.
18. Discuss heart failure.
19. List the risk factors involved in heart attacks.
20. Explain why inadequate blood supply, anatomical disorders, and malfunctions of conduction are primary reasons for heart trouble.
21. Contrast patent ductus arteriosus, septal defects, and valvular stenosis as congenital heart defects.
22. List the four abnormalities of the heart present in tetralogy of Fallot.
23. Define atrioventricular block, atrial flutter, atrial fibrillation, and ventricular fibrillation as abnormalities of the conduction system of the heart.
24. List the causes and symptoms of aneurysms, atherosclerosis, and hypertension.
25. Discuss the diagnosis of atherosclerosis by the use of angiography.
26. Define key medical terms associated with the heart and blood vessels.

(pages 298–299)   **A.** Location, size, and shape

1. What is the main function of the heart?

2. Closely examine Figure 12-1 in your text. Consider the location, size, and
   shape of your heart as you do this exercise. Trace its outline on your body.

   a. Your heart lies in the _____ portion of your thorax,

      between your two _____. About (*one third? one-
      half? two-thirds?*) of the mass of your heart lies to the left of the midline
      of your body.

   b. Your heart is about the size and shape of your _____.

   c. The pointed part of your heart, called the _____, lies

      in the _____ intercostal space, about _____ cm

      (_____ inches) from the midline of your body.

   d. The base of your heart is located just inferior to your

      _____ ribs.

3. What structures protect the heart?
   a. Anteriorly

   b. Posteriorly

   c. Inferiorly

(pages 299–305)   **B.** Pericardium, walls and chambers, great vessels, valves

1. List in order (from most superficial to deepest) the coverings over the heart.
   Briefly describe the structure and functions of each.

2. What is the function of *pericardial fluid?*

3. Contrast according to structure and location:
   a. Epicardium

   b. Myocardium

   c. Endocardium

4. Refer to Figure LG 12-1 as you do this exercise about heart structure. |3|

   a. The heart is divided into _____ spaces or chambers.

   b. The superior ones are known as _____ . Each of

   these has an appendage, or _____ . Most of the lin-
   ing of the atria is (*smooth? ridged?*). The tissue separating the two atria is

   named the _____ .

   c. The two inferior chambers of the heart are called the _____

   _____. Their walls are (*thicker? thinner?*) than
   those of the atria. The two ventricles are separated by the

   _____ . Externally, a groove, known as the

   _____ , marks the location of the septum. (See Figure 12-2 in the text.)

   d. The atrial myocardium is separated from that of the ventricles by the

   "cardiac skeleton" composed of _____ , which also

   forms _____ there.

   e. Blood from all parts of the body except the lungs flows into the chamber

   named the _____ . Blood from superior body parts

   enters the heart via the vein called the _____ . The

150

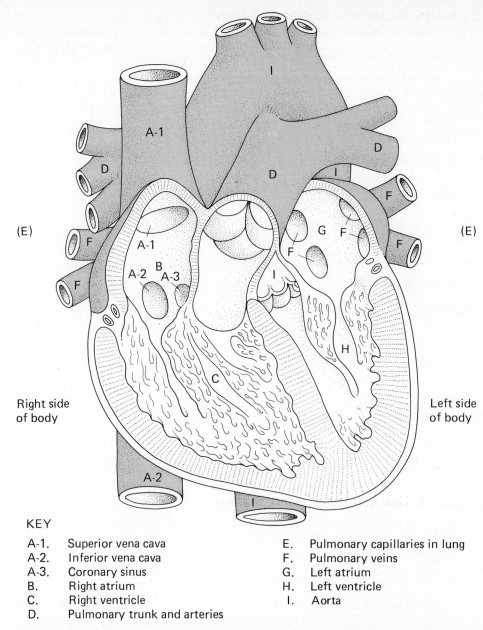

KEY

A-1. Superior vena cava
A-2. Inferior vena cava
A-3. Coronary sinus
B.   Right atrium
C.   Right ventricle
D.   Pulmonary trunk and arteries

E.   Pulmonary capillaries in lung
F.   Pulmonary veins
G.   Left atrium
H.   Left ventricle
I.   Aorta

Figure LG 12-1 Structure of the heart, anterior internal view. Key letters are arranged alphabetically along the path of blood through the heart. Complete as indicated.

inferior vena cava returns blood from _____ . Blood from vessels supplying heart tissue returns to the right atrium via the vessel named the _____ .

f. Blood in the right atrium passes into the _____ . From here it is squeezed into a vessel named the _____ which leads to the _____ .

g. Oxygenated blood returns to the heart by means of the pulmonary _____ . There are _____ of these, two from each lung. They empty into the _____ of the heart.

h. Blood in the left atrium flows into the _____ . This
chamber is the (*thickest? thinnest?*) chamber in the heart since it
must squeeze blood into the _____ , which gives
branches leading to all body parts (except the lungs).

5. On Figure LG 12-1:
   a. Draw arrows indicating direction of blood flow.
   b. Color red the structures that contain highly oxygenated blood; color blue
      the structures containing blood which is low in oxygen.
   c. Indicate the location of and label the four valves which control blood
      flow through the heart.

6. State the general function of valves. Describe the structure of valves.

7. Check your understanding of the valves by doing this matching exercise.
   More than one answer may be required for each description.

_____ a. Also called the mitral valve

_____ b. An atrioventricular (AV) valve

_____ c. Prevents backflow of blood into
         the right ventricle

_____ d. Has half-moon–shaped leaflets,
         or cusps

A. Aortic semilunar
B. Bicuspid
P. Pulmonary semilunar
T. Tricuspid

8. Describe according to structure and function:
   a. Chordae tendineae

   b. Papillary muscles

C. Conduction system, cardiac cycle, sounds, control (pages 305–310)

1. Describe the conduction system of the heart according to:
   a. Function

151

   b. Type of tissue

2. Name in order the structures which make up the conduction system. Then draw and label these on Figure LG 12-1.

5

3. What structural feature of the heart makes the AV node and the bundle of His necessary for conduction to the entire heart?

4. Define *electrocardiogram* (*ECG*) and state its significance.

6

5. Complete this exercise about the cardiac cycle and heart sounds.

   a. Define *cardiac cycle*.

   b. The term _____ refers to contraction of heart muscle, while the term _____ means heart relaxation.

   c. At a heart rate of 72 beats/min, one cardiac cycle requires _____ sec. During the first 0.1 sec of the cycle, the atria are in (*systole? diastole?*), while the ventricles are in (*systole? diastole?*).

   d. Next the ventricles contract, causing the atrioventricular valves to (*open? close?*). Soon afterwards a buildup of pressure within the ventricles forces the semilunar valves to (*open? close?*). Ventricular systole lasts

      _____ sec during which blood is squeezed out of the heart and into the great vessels.

e. During the last half of the cardiac cycle (_____ sec), all heart muscle is relaxed. So both the atria and the ventricles are said to be in (*systole? diastole?*). At first all valves are closed, but then increased filling of atria with blood from the great veins causes the (*atrioventricular? semilunar?*) valves to open. Blood can again enter the ventricles and the cycle is repeated.

f. In general, what causes the heart sounds that occur during each cardiac cycle?

g. The first heart sound (*lubb*) is associated with closure of the

_____ valves, while the second sound (*dupp*)

occurs as a result of closure of the _____ valves.

6. What causes *heart murmurs*?

7. Study Figure 12-5 in your text, and then determine on a skeleton and on yourself the surface projections of heart valves and the sites where heart sounds may best be heard.

8. Complete the following exercise describing control of heart rate and its effect on blood pressure.

   7

   a. Cardiac output is directly related to blood pressure (BP). This means that when the heart increases its output of blood, more blood is present in vessels, exerting (*increased? decreased?*) BP. Cardiac output must be kept in check, however, or BP would skyrocket.

   b. Cardiac and blood pressure control centers are located in the

   _____ of the brain. Sensory nerves must constantly inform the medulla of the need for adjustments. Cranial nerves IX and X carry these messages from receptors sensitive to BP. These pressorecep-

   tors are located in the walls of the _____ and the

   _____.

   c. When pressoreceptors indicate that BP is increased slightly, in order to maintain homeostasis BP must be caused to (*decrease slightly? increase even more?*). This can be done by (*increasing? decreasing?*) heart rate.

   d. Consider the cardiac center as comparable to the control center of an automobile. The center has an accelerator portion (like a gas pedal) and

   an inhibitor portion (like the _____ of a car).

   e. When BP is reported to be too high (as if the car is running too fast), the cardioacceleratory center must be (*stimulated? inhibited?*); in the car analogy, you would (*step on the gas? take your foot off the gas?*). At the same time the cardioinhibitory center (brakes) would be (*stimulated? inhibited?*).

   f. Dual control is carried out by autonomic nerve fibers. Sympathetic (accelerator) nerve messages would (*increase? decrease?*), while parasym-

   pathetic nerve impulses along the _____ nerve to the heart would (*increase? decrease?*).

g. As a result of these nerve messages heart rate would (*increase? decrease?*) and thereby (*raise? lower?*) BP in response to each occasion of slightly increased BP. This is a statement of _____ law. The mechanism is an example of (*positive? negative?*) feedback.

9. The Bainbridge reflex works by a (*positive? negative?*) feedback mechanism. Describe this reflex.

(pages 310–316) **D. Blood vessels**

⁸ 1. Trace the route of a drop of blood.

Aorta ⟶ artery ⟶ _____ ⟶

_____ ⟶ _____

⟶ _____ ⟶ heart

2. Label the layers of the blood vessels in Figure LG 12-2. Notice the relative thickness of each layer in the different vessels.

⁹ 3. State the two major properties of arteries.

Explain what structural features account for these properties.

4. Contrast *vasoconstriction* with *vasodilation*.

5. Define *anastomosis*.

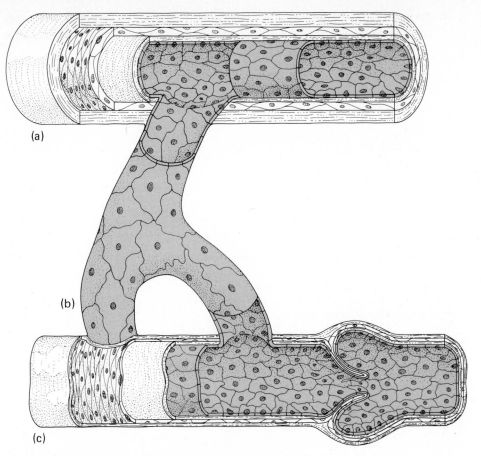

Figure LG 12-2 Structure of blood vessels. (a) Artery. (b) Capillary. (c) Vein. The relative size of the capillary is enlarged for emphasis.

List several places where anastomoses are located. ☐10

6. Describe the gradual changes in structure of an arteriole in its course between an artery and a capillary.

7. List two principal functions of arterioles.

8. Explain how the distribution of capillaries in the body correlates to the activity of the tissue. Then name areas where you would expect capillary supply to be extensive, limited, or absent.

9. Discuss how the structure of capillaries is suited to their function.

10. Define:
   a. Precapillary sphincters

   b. Sinusoids

11. Complete this exercise about structure of veins.

   a. Veins have (*thicker? thinner?*) walls than arteries. This structural feature relates to the fact that the pressure in veins is (*more? less?*) than in arteries. The pressure difference is demonstrated when a vein is cut; blood leaves a cut vein in (*rapid spurts? an even flow?*).
   b. Gravity exerts back pressure on blood in veins located inferior to the heart. To counteract this, veins contain _____ .
   c. When valves weaken, veins become enlarged and twisted. This condition is called _____ . This occurs more often in (*superficial? deep?*) veins. Why?

   d. Varicosities of the anal canal are known as _____ .
   These may be brought on by straining during _____
   or _____ .

12. Name two parts of the body where vascular (venous) sinuses are located.

1. Study Figure 12-8 in your textbook. Then describe the general plan of systemic circulation. Remember that this is a "closed circuit," that is, all blood vessels are continuous from one side of the heart back to the other side of the heart.

2. Use Figures LG 12-3 and LG 12-4 to complete this exercise about arteries.
   a. The major artery from which all systemic arteries branch is the

      _____. It exits from the _____ of the heart. The aorta has three portions. Label these on Figure LG 12-3.

   b. The first arteries to branch off of the aorta are the _____

      arteries, supplying the _____. Label these on the figure.

   c. The arch of the aorta has three vessels branching from it. Label these in order, beginning with the vessel closest to the heart.

   d. Locate and label the two branches of the brachiocephalic artery. The

      right subclavian artery supplies the _____; the right

      common carotid supplies the _____.

   e. List the names given to the subclavian artery as it passes from the thorax

      into the arm: _____. Label these on the figure. Then continue the drawing of major arteries in the forearm and hand. Label them. (Check Figure 12-11 in the text.)

   f. Besides supplying the arm, the right subclavian artery also sends a branch that ascends the neck through foramina in the cervical vertebrae.

      This is the _____ artery. Label it. This artery termi-

      nates by forming the _____ artery.

   g. Again locate the right common carotid artery. At what point does it bi-

      furcate (divide into two branches)? _____
      Label its two branches, listed below. In general, what does each supply? (See Exhibit 12-4.)
      Right external carotid artery:

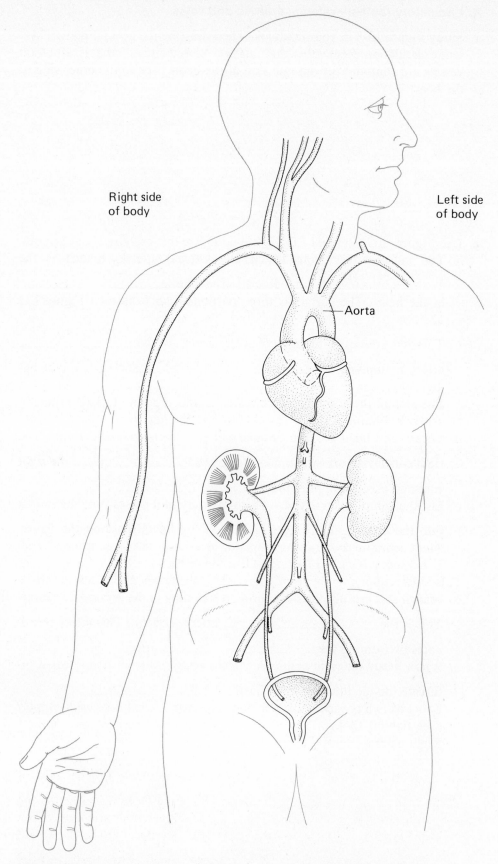

Right side
of body

Left side
of body

Aorta

Figure LG 12-3 Major arterial branches from aorta. Complete figure as directed.

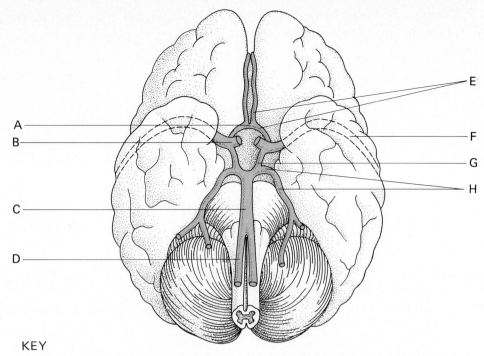

Figure LG 12-4 Arterial supply to the brain including vessels of circle of Willis. View is from undersurface of the brain. Numbers in parentheses in key indicate whether arteries are single (1) or paired (2).

Right internal carotid artery:

h. Blood supply to the brain is vital. An anastomosis, called the circle of

_____ (Figure LG 12-4), is located just below the brain. Its many branches supply blood to cerebral arteries. Which two vessels provide blood to the anterior of the circle?

_____ Which supply the posterior portion of the

circle? _____ Identify these vessels on Figure LG 12-4.

3. Contrast the *visceral branches* of aorta with the *parietal branches* of aorta. Then list several examples of each.

4. Complete the table describing visceral branches of the abdominal aorta. Note that arteries are listed in the order in which they emerge from the aorta (from superior to inferior). Label these on Figure LG 12-3. (See also Figure LG 22-1, page 327).

Artery	Structures Supplied
a. Celiac (three major branches) 1. 2. 3.	1. Liver 2. 3.
b.	Small intestine and part of large intestine
c.	Adrenal (suprarenal) glands
d. Renal	
e. Inferior mesenteric	
f.	Testes (or ovaries)

5. The aorta ends at about level _____ vertebra by dividing into right and left

_____ arteries. Label these on Figure LG 12-3.

6. Label the two branches of each common iliac artery. Name the structures supplied by each. (See Exhibit 12-10.)
   a. Internal iliac artery

b. External iliac artery

7. Trace three possible routes of a drop of blood from the common iliac artery to the little toe. (Refer to Figure 12-13 in your textbook.) The first pathway is started for you. Common iliac artery ⟶ external iliac artery ⟶

8. Name the three main vessels that empty venous blood into the right atrium of the heart.

12

What are the major regions of the body drained by each?

9. Contrast the structure of cranial venous sinuses with that of "true veins." Describe the venous pathways draining the brain.

10. Once you are familiar with the pathways of systemic arteries, you know much about the veins which accompany these arteries. Demonstrate this by tracing routes of a drop of blood along the following deep vein pathways. Name the vessels in order.

13

   a. From the thumb side of the left forearm to the right atrium

   b. From the medial aspect of the left knee to the right atrium

   c. From the right kidney to the right atrium

11. Contrast veins with arteries in this exercise.
   a. Veins have (*higher? lower?*) blood pressure than arteries and a (*faster? slower?*) rate of blood flow.
   b. Veins are therefore (*more? less?*) numerous, in order to compensate for slower blood flow in each vein. The "extra" veins consist of vessels located just under skin (in subcutaneous fascia); these are known as

   _____ veins.
12. Now trace a drop of blood through these superficial pathways. 15

   a. From the skin on the thumb side of the left forearm to the right atrium

   b. From the skin on the medial aspect of the left knee to the right atrium

13. On Figures 12-16 and 12-18 in your text, locate the following veins. Identify the location where each empties into a deep vein. (Consult Exhibits 12-9 and 12-12 in your text for help.)
   a. Basilic

   b. Median cubital

   c. Small saphenous

14. Veins connecting the superior and inferior venae cavae are named

_____ veins. Identify the azygos system of veins on Figure 12-11 in the text. What parts of the thorax do they drain? Explain how these veins may help to drain blood from the lower part of the body if the inferior vena cava is obstructed.

15. Name the visceral veins that empty into the abdominal portion of the inferior vena cava. Then color them blue on Figure LG 22-1. Contrast accompanying systemic arteries by coloring them red on that figure.

(pages 340–345) **F.** Circulatory routes: coronary, hepatic portal, pulmonary, and fetal circulation

1. Defend or dispute this statement: "The myocardium receives all of the oxygen and nutrients it needs from blood which is passing through its four chambers."

2. The arteries which supply heart tissue are named the _____

_____ arteries. They are branches from the (*ascending? arch? descending?*) portion of the aorta.

3. Describe the pathway of blood supplying heart tissue. Name the types of vessels in order from coronary arteries back to the right atrium.

4. Match the coronary artery branches with descriptions given. More than one answer may be required. 16

_____ a. Supplies the left ventricle

_____ b. Supplies both left atrium and left ventricle

_____ c. Major branch of the right coronary artery

_____ d. Located on anterior surface of the heart in a groove between the ventricles

A. Anterior interventricular
C. Circumflex
M. Marginal
P. Posterior interventricular

5. Define _ischemia_ and explain how it is related to _angina pectoris_.

How does nitroglycerin help to relieve pain associated with angina pectoris?

6. What is the technical term for a "heart attack"? What does the term mean and what might cause this event?

7. Why are serum enzyme levels (such as for CPK, LDH, and SGOT) studied in patients suspected of having heart attacks?

8. Note on Figure LG 22-1 that blood from digestive organs (such as blood in superior and inferior mesenteric veins) does *not* drain directly into the inferior vena cava. Complete this exercise about this blood route.

17

a. Virtually all blood from digestive organs, namely, _____

_____ ,

as well as blood from the spleen, empties into veins which lead to the

single _____ vein. This vessel enters the undersurface of the liver.

b. Look at Figure 12-20 in the text and name the major veins which drain into this large hepatic portal vein.

c. Note the uniqueness of this vein: it carries (*port* = carry) blood from a set of organs to another organ. Other veins of the body empty into

_____ .

d. You have studied another vessel which enters the liver. This is the

_____ artery, which carries oxygenated blood. (See Figure LG 21-4.)

e. In the liver blood from both the hepatic artery and the portal vein mixes (see Figure 21-15 in the text) as it travels through tortuous capillarylike

structures named _____ .

f. Blood leaves the liver via the right and left _____ ,
which exit from the top of the liver and empty into the inferior vena cava.

g. What functions are served by the special hepatic portal pathway?

9. Write a sentence describing pulmonary circulation. (Review the pathway using Figure LG 12-1.)

10. What aspects of fetal life require the fetus to possess special cardiovascular structures not needed after birth?

11. Complete the following table about fetal structures. (Note that the structures are arranged in order of blood flow from fetal aorta back to fetal aorta.)

Fetal Structure	Structures Connected	Function	Fate of Structure After Birth
a.		Carries fetal blood low in oxygen and nutrients, high in wastes	
b. Placenta	(Omit)		
c.	Placenta to liver and ductus venosus		
d.		Branch of umbilical vein, bypasses liver	
e. Foramen ovale			
f. Ductus arteriosus			

18

12. For each of the following pairs of fetal vessels, circle the vessel with higher oxygen content.

 a. Umbilical artery/umbilical vein
 b. Ductus arteriosus/ductus venosus
 c. Femoral artery/femoral vein
 d. Pulmonary artery/pulmonary vein
 e. Aorta/thoracic portion of inferior vena cava

(pages 345–352)   **G. Applications to health; key medical terms**

1. List five risk factors related to heart attacks. Explain why each factor increases the likelihood of an attack.

2. Define each of the following terms.
 a. Arrhythmia

 b. Cardiac arrest

 c. Patent ductus arteriosus

 d. Septal defect

e. Cyanosis

f. Valvular stenosis

g. Tetralogy of Fallot

h. "Blue baby"

3. Define *A V block.* Discuss the different degrees of AV block.

4. Contrast *atrial fibrillation* with *ventricular fibrillation.*

5. Define *aneurysm.* Which artery is most susceptible to aneurysm?

6. Contrast *arteriosclerosis* and *atherosclerosis*.

7. List three reasons why atherosclerosis may be life-endangering.

8. Describe the development of atherosclerosis. List risk factors related to this disease.

9. How are *angiograms* used to diagnose arteriosclerosis?

10. Define *hypertension*. What clinical measurement determines the diagnosis of the condition?

11. Contrast *primary hypertension* and *secondary hypertension*.

12. Define each of the following key medical terms.
    a. Cyanosis

    b. Cardiac arrest

    c. Catheterization

    d. Cor pulmonale

    e. Defibrillator

    f. Hematoma

g. Occlusion

h. Phlebitis

Answers to Numbered Questions in Learning Activities

1. (a) Mediastinal, lungs, two-thirds. (b) Fist. (c) Apex, fifth, 8, 3. (d) Second.

2. (a) Sternum, costal cartilages. (b) Bodies of vertebrae, ribs. (c) Diaphragm.

3. (a) 4. (b) Atria, auricle, smooth, interatrial septum. (c) Ventricles, thicker, interventricular septum, sulcus. (d) Connective tissue, valves. (e) Right atrium, superior vena cava, parts of the body inferior to the heart, coronary sinus. (f) Right ventricle, pulmonary trunk, pulmonary arteries and smaller vessels in lungs. (g) Veins, 4, left atrium. (h) Left ventricle, thickest, aorta.

4. (a) B. (b) B, T. (c) P. (d) A, P.

5. The fibrous cardiac skeleton completely separates atrial myocardium from ventricular myocardium.

6. (a) A complete heart beat. (b) Systole, diastole. (c) 0.8, systole, diastole. (d) Close, open, 0.3. (e) 0.4, diastole, atrioventricular. (f) Turbulence in blood flow created by closure of valves. (g) Atrioventricular, semilunar valves.

7. (a) Increased. (b) Medulla, carotid sinus, and aorta. (c) Decrease slightly, decreasing. (d) Brakes. (e) Inhibited, take your foot off the gas, stimulated. (f) Decrease, vagus, increase. (g) Decrease, lower, Marey's, negative.

8. Arteriole ⟶ capillary ⟶ venule ⟶ vein.

9. Elasticity, contractility.

10. Coronary arteries, circle of Willis at the base of the brain, in palm and foot, and around major joints.

11. (a) Thinner, less, an even flow. (b) Valves. (c) Varicose veins, superficial, skeletal muscles around deep veins prevent overstretching. (d) Hemorrhoids, constipation, childbirth.

12. Superior vena cava, inferior vena cava, and coronary sinus.

13

The Lymphatic System

The lymphatic system works side by side with the cardiovascular system in transporting fluids. It also provides vital defense mechanisms. In this chapter you will examine components, structure, and functions of the lymphatic system and follow circulatory pathways of lymph (Objectives 1–6). You will learn how lymph circulation is maintained (7) and how edema develops (8). You will also learn medical terminology associated with the lymphatic system and immunity (9).

Topics Summary

A. Lymphatic vessels, structure of lymph nodes, lymph circulation, principal groups of lymph nodes

B. Lymphatic organs

C. Key medical terms associated with the lymphatic system.

Objectives

1. Define the lymphatic system and list its functions.

2. Describe the structure and origin of lymphatics and contrast them with veins.

3. Describe the histological aspects of lymph nodes and explain their functions.

4. Trace the general plan of lymph circulation from lymphatics into the thoracic duct or right lymphatic duct.

5. Describe the principal lymph nodes of the head and neck, extremities, and trunk, their location, and the areas they drain.

6. Contrast the locations and functions of the tonsils, spleen, and thymus gland as lymphatic organs.

7. Explain the forces responsible for maintaining the circulation of lymph.

8. Discuss how edema develops.

9. Define key medical terms associated with the lymphatic system.

(pages 356–370)   **A. Lymphatic vessels, structure of lymph nodes, lymph circulation, principal groups of lymph nodes**

1. List the components of the lymphatic system.

2. What is the primary function of this system? Name three additional functions of this system.

3. Contrast in terms of structure:
   a. Blood capillaries/lymph capillaries

   b. Veins/lymphatics

4. What is a *lymphangiogram?* Explain how it is used clinically.

5. Refer to Figure LG 13-1, showing a lymph node. Label: *capsule, hilus, lymph nodules, medullary cords, afferent lymphatic vessels, cortical sinuses, medullary sinuses,* and *efferent lymphatic vessels.* State the actual size range of lymph nodes. Draw arrows to show the pathway of lymph.

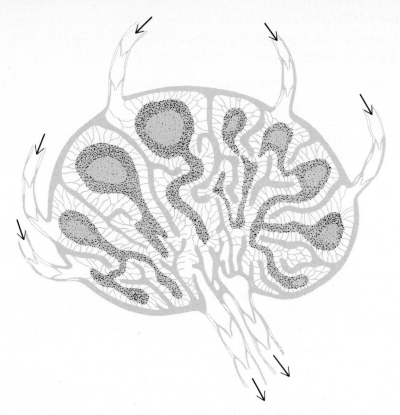

Figure LG 13-1 Structure of a lymph node. Label according to directions.

6. Where are lymphocytes produced?

1

7. Describe how lymph is "processed" as it circulates through lymph nodes. How does this function of lymph nodes explain the fact that nodes ("glands") become enlarged and tender during infections?

8. Describe the general pathway of lymph circulation from fluids in plasma and interstitial areas through lymph vessels and nodes to major collecting vessels and back to blood plasma. Name the structures along this route.

9. Name the areas of the body drained by each of the following lymphatic trunks.
   a. Lumbar

   b. Intestinal

   c. Jugular

   d. Subclavian

   e. Bronchomediastinal

10. Describe the *cisterna chyli* according to location, structure, and function.

11. Answer the following questions about the largest lymphatics.
    a. In general, the thoracic duct drains (*three-fourths? one-half? one-fourth?*) of the body's lymph.

    b. The remaining lymph flows into the _____ . The regions of the body from which this large lymphatic collects lymph are

       _____ .

    c. The thoracic duct empties into the _____ ; the right

       lymphatic duct drains into the _____ .

    d. Lymph fluid starts out originally in plasma and circulates as interstitial fluid. Then as lymph it undergoes an extensive cleaning as it passes through nodes. Finally this fluid in lymph is returned to

       _____ in the subclavian veins.

12. Examine Exhibits 13-1 to 13-5 and related figures in your textbook. Then match the major lymph nodes listed with the descriptions.

a. These nodes directly drain the knee.

   _____

b. These nodes would be most likely to enlarge as defense against a serious infection following a cut on the little

   finger.   _____

c. Infection of the skin of the leg or external genitals would cause these

   nodes to become swollen.   _____

d. These drain the deep lymphatics of

   the lower extremity.   _____

e. During infections of the nasal cavity or nasopharynx, these are likely to swell. They are the largest group of

   nodes in the neck.   _____

f. Lymph from infected bladder or uterus (pelvic viscera) would pass first

   into these nodes.   _____

g. Located adjacent to the abdominal aorta, they receive efferents from common iliacs and also drain kid-

   neys.   _____

h. These are visceral nodes that drain the stomach, liver, spleen, and pancreas, organs supplied by celiac ar-

   tery.   _____

i. These nodes lie along the superior mesenteric artery and drain structures

   supplied by this artery.   _____

j. These nodes drain and provide first defense against infections of the lower respiratory passageways and the

   lungs.   _____

Celiac
Deep cervical
Deep inguinal
External, internal, and common iliacs
Lumbar
Popliteal
Superior mesenteric
Superficial inguinal
Supratrochlear
Tracheal, bronchial, and pulmonary

13. Take particular note of the widespread drainage of the breasts (Figures 13-7b and 13-10 in the text). Discuss how this arrangement might cause breast cancer to spread (metastasize) to other sites, such as the pectoralis muscles, liver, lungs, and opposite breast.

(pages 371–373)   **B.** Lymphatic organs

1. Answer these questions about lymphatic organs.
   a. Name three lymphatic organs.

   b. State locations of the three masses of tonsillar tissue.

   c. In which areas would severe blows most likely cause injury to the spleen, possibly necessitating a splenectomy?

   d. Contrast *white pulp* with *red pulp* of the spleen.

   e. List five functions of the spleen.

14

Nervous Tissue

The diversity and complexity of movements of the musculoskeletal and circulatory systems point to the necessity for communication and control in the body. In the next six chapters you will learn about the mechanisms that provide such control over all body systems through nerve impulses and hormonal messages.

In this chapter you will study functions and divisions of the nervous system (Objective 1), and you will consider the cells that make up nervous tissue (2–4). Then you will learn how nerve impulses take place, including their transmission across synapses (5–6).

Topics Summary

A. Organization

B. Histology

C. Physiology: regeneration, nerve impulse

Objectives

1. Classify the organs of the nervous system into central and peripheral divisions.

2. Contrast the histological characteristics and functions of neuroglia and neurons.

3. Classify neurons by shape and function.

4. List the necessary conditions for the regeneration of nervous tissue.

5. Define a nerve impulse and describe its importance in the body.

6. Explain how a nerve impulse is transmitted across a synapse.

Learning Activities

(pages 376–377)   **A. Organization**

1. Defend or dispute this statement: "It is impossible for humans to live without a functioning nervous system."

2. Contrast the roles of the *nervous* and *endocrine* systems in maintenance of homeostasis.

3. Write an outline summarizing the divisions of the nervous system.

4. Check your understanding of nervous system organization by doing this exercise.

    a. The central nervous system (abbreviated _____) consists of two structures. These are the _____ and the _____

    _____ .

    b. Another name for *afferent* is (*sensory? motor?*).

    c. Efferent nerves pass from the _____ to _____

    _____ and _____ . Efferent nerves are often referred to as (*sensory? motor?*).

    d. Efferent nerves to skeletal (voluntary) muscles are called _____

    _____ efferent nerves.

    e. The _____ nervous system (abbreviated _____) consists of efferent nerves that innervate three types of tissue. These

    are _____ , _____ , and

    _____ .

    f. The two divisions of the autonomic nervous system (ANS) are

    _____ and _____ .

1. What two kinds of cells make up the nervous system? State general functions of each.

2. Complete the table describing neuroglia cells.

Type	Description	Functions
a.	Star-shaped cells	
b. Oligodendroglia		
c.		Phagocytic

3. Draw a diagram of a neuron. Label: *cell body, nucleus, axon, dendrites, mitochondria,* and *neurofibrils.*

4. How are *Nissl* bodies important in the function of a neuron?

5. Examine the enlargement of a cross section and longitudinal section of a myelinated nerve fiber in Figure LG 14-1. Label: *neurofibrils, myelin sheath, neurilemma* (with *Schwann cells*), and *node of Ranvier.*
6. How does myelination relate to the color of:
   a. White matter

2

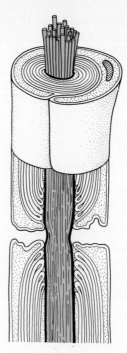

Figure LG 14-1 Diagram of nerve fiber to be labeled as directed.

    b. Gray matter

3   7. Explain how myelin is laid down on nerve fibers in the:
    a. PNS

    b. CNS

8. Name and briefly describe three types of neurons based on structural characteristics and three types of neurons classified according to functional differences.

9. Identify phrases that describe a *neuron* and those that describe a *nerve*. 4

a. A single nerve cell; consists of a cell body with axon and dendrites:

_____

b. Bundle of axons and dendrites of many neurons, both afferent and efferent, somatic and autonomic; contains no cell bodies:

_____

c. Located entirely outside of the CNS and in the PNS; macroscopic in diameter: _____

d. Located partly in CNS and partly in PNS; microscopic in diameter:

_____

10. Identify the type of nerve fiber that transmits each of the kinds of nerve impulses listed below. 5

_____ a. All ANS fibers are of this type.

_____ b. Pain from a thorn in your skin is sensed via fibers of this type.

_____ c. Pain from a spasm of smooth muscle in the gallbladder is sensed by means of fibers of this type.

_____ d. With your eyes closed, you can tell the position of your skeletal muscles and joints due to nerve impulses which pass along this type of fiber.

_____ e. In order to increase your heart rate, impulses pass from your brain to your heart via this type of nerve fiber.

_____ f. These nerve fibers carry impulses from the CNS to muscles in your fingers used in writing.

GSA. General somatic afferent

GSE. General somatic efferent

GVA. General visceral afferent

GVE. General visceral efferent

11. Contrast the structure of a typical afferent (sensory) neuron (Figure 14-4a) with that of a typical efferent (motor) neuron (Figure 14-2a). Draw a diagram of each of these types of neurons. Label *dendrite (s)*, *cell body,* and *axon.*

C. Physiology: regeneration, nerve impulse

    1. Which of the following can regenerate if destroyed? Explain why in each case.
      a. Neuron cell body

      b. CNS nerve fiber

      c. PNS nerve fiber

    2. State two characteristics of neurons that make it possible for them to carry nerve impulses.

      6

      Define each of those characteristics.

    3. Myelin (*increases? decreases?*) the rate of nerve transmission.

    4. Examine Figure 14-5 in your text. Then write brief descriptions of each of the following parts of a synapse.
      a. Presynaptic neuron

      b. Presynaptic knob

15

The Spinal Cord and the Spinal Nerves

In this chapter you will begin to see the organization of nervous tissue into functional units, exemplified by the spinal cord and nerves. You will define important terms related to the cord and nerves (Objective 1). You will describe principal features of the spinal cord (2–4, 6), study reflexes (5, 7–8), and examine the structure, distribution, and function of spinal nerves (9–14). You will also consider injuries and disorders involving the spinal cord and nerves (15–18).

Topics Summary

A. Grouping of neural tissue

B. Spinal cord: general features, protection and coverings, structures, conduction pathway

C. Spinal cord: reflex center, reflex arc

D. Spinal nerves

E. Applications to health

Objectives

1. Define white matter, gray matter, nerves, ganglion, tract, nucleus, and horn.
2. Describe the gross anatomical features of the spinal cord.
3. Explain how the spinal cord is protected by the meninges.
4. Describe the structure of the spinal cord in cross section.
5. Explain the functions of the spinal cord as a conduction pathway and a reflex center.
6. List the location, origin, termination, and function of the principal ascending and descending tracts of the spinal cord.
7. Describe the components of a reflex arc.
8. Compare the functional anatomy of a stretch reflex, flexor reflex, and crossed extensor reflex.
9. Describe the composition and coverings of a spinal nerve.
10. Name the 31 pairs of spinal nerves.
11. Explain how a spinal nerve branches upon leaving the intervertebral foramen.
12. Explain the composition and distribution of the cervical, brachial, lumbar, sacral, and coccygeal plexuses.
13. Define an intercostal nerve.
14. Define a dermatome and its clinical importance.
15. Describe spinal cord injury and list the immediate and long-range effects.
16. Identify the effects of peripheral nerve damage and conditions necessary for its regeneration.
17. Distinguish between sciatica and neuritis.
18. Explain the cause and symptoms of shingles.

A. Grouping of neural tissue

1. 1. Complete the table about structures in the nervous system.

Structure	Color (Gray or White)	Composition (Cell Bodies or Nerve Fibers)	Location (CNS or PNS)
a. Nerve			
b. Tract			
c. Ganglia			
d. Nucleus			

2. Refer to Figure 15-3a in your text. Contrast *gray horns* with *white columns* of the spinal cord.

(pages 386–392) **B.** Spinal cord: general features, protection and coverings, structure, conduction pathway

1. Describe the spinal cord according to:
   a. Shape

   b. Location, length; estimate its extent on a partner.

   c. Enlargements

2. Match the names of the structures with descriptions given.

_____ a. Tapering inferior end of spinal cord

_____ b. Any region of spinal cord from which one pair of spinal nerves arises

_____ c. Nonnervous extension of pia mater; anchors cord in place

_____ d. "Horse's tail"; extension of spinal nerve roots in lumbar and sacral regions within subarachnoid space

Ca. Cauda equina
Co. Conus medullaris
F. Filum terminale
S. Spinal segment

3. In what ways is the spinal cord protected? List four structures or other factors that are protective.

4. Check your understanding of the protective coverings over the spinal cord by labeling A–F on Figure LG 15-1 using the following terms: _arachnoid, brain or spinal cord tissue, dura mater, pia mater, subarachnoid space,_ and _subdural space._

3

5. Now color the following structures on Figure LG 15-1: dura mater, blue; arachnoid, yellow; pia mater, green; blood vessels, red.

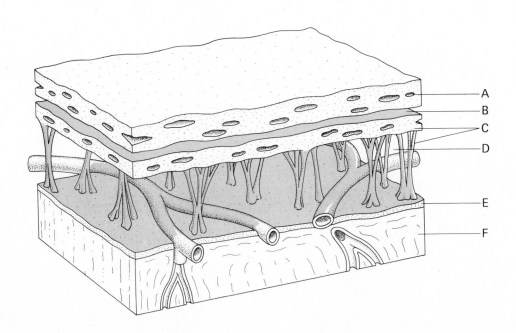

— A
— B
— C
— D
— E
— F

Figure LG 15-1 Meninges. Color and label as directed.

4 6. Do the following exercise about spinal (lumbar) puncture.
     a. At what level of the vertebral column is the needle inserted?

     b. Why is the puncture done here rather than at level L1 to L2?

     c. The needle enters the (*subarachnoid? subdural? epidural?*) space where
        cerebrospinal fluid is located.
  7. Label the following structures on Figure LG 15-2a: *anterior median fissure,
     posterior median sulcus, gray commissure, central canal,* all gray *horns,* and
     all white *funiculi.* Color gray (shade with pencil) all areas that consist of
     gray matter.

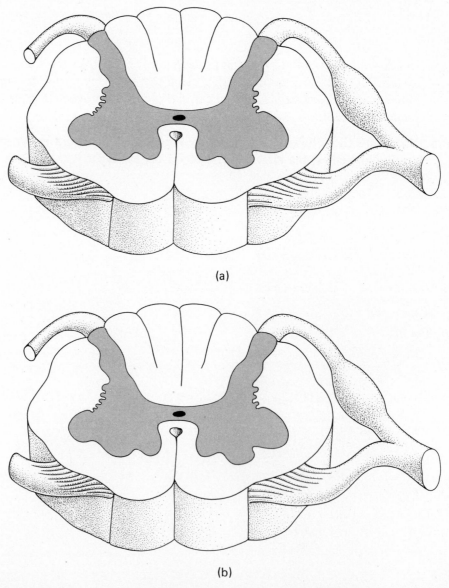

(a)

(b)

Figure LG 15-2 Outline of spinal cord, roots, nerves. Complete diagram as directed.

8. Circle or fill in the correct answers about tracts.

   a. Ascending tracts are all (*sensory? motor?*).

   b. All motor tracts in the cord are (*ascending? descending?*).

   c. The name *lateral corticospinal* indicates that the tract is located in the

   _____ funiculus, that it originates in the (*cerebral cortex? thalamus? spinal cord?*), and that it ends in the _____ .

   d. The lateral corticospinal tract is (*ascending, sensory? descending, motor?*).

9. Draw and label the following tracts on Figure LG 15-2b: *anterior corticospinal, anterior spinothalamic, fasciculus cuneatus, fasciculus gracilis, lateral corticospinal,* and *lateral spinothalamic.* Color ascending (sensory) tracts red, descending (motor) tracts blue.

10. For an additional challenge, label these tracts on Figure LG 15-2b and color them as described above (optional): *anterior spinocerebellar, posterior spinocerebellar, rubrospinal, tectospinal,* and *vestibulospinal.*

11. Match the names of the tracts with the descriptions given.

_____ a. Conveys impulses that tell you that you touched a hot stove

_____ b. Allow you to be conscious of the position of your body parts; help you to assess weight and shape of objects

_____ c. Located in anterior white column; starts in spinal cord and ends in thalamus

_____ d. Sends impulses from your brain to enable you to move your skeletal muscles voluntarily

AS. Anterior spinothalamic
LS. Lateral spinothalamic
C. Anterior and lateral corticospinal
F. Fasciculus cuneatus and fasciculus gracilis

C. Spinal cord: reflex center, reflex arc   **(pages 392–398)**

1. State the two major functions of the spinal cord.

2. Define each of these terms. Be sure your definitions distinguish them from one another.
   a. Reflex

   b. Reflex arc

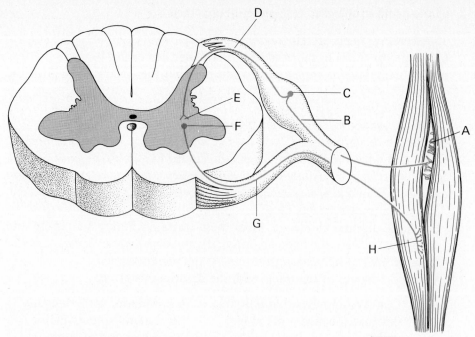

Figure LG 15-3 Reflex arc: stretch reflex. Letters refer to question 7

c. Spinal reflex

d. Somatic reflex

e. Visceral reflex

3. Label structures A–H on Figure LG 15-3, using the following terms: *effector, motor neuron axon, motor neuron cell body, receptor, sensory neuron axon, sensory neuron cell body, sensory neuron dendrite,* and *synapse.* Note that these structures are lettered in alphabetical order along the conduction pathway of a reflex arc.

7

8 4. Answer these questions about the conduction pathway in Figure LG 15-3.

a. How many neurons does this reflex contain? _____ The neuron that conveys the impulse toward the spinal cord is a _____ neuron; the one that carries the impulses toward the effector is a _____ neuron.

b. This is a (*mono? poly?*)-synaptic reflex arc. The synapse, like all somatic synapses, is located in the (*CNS? PNS?*).

c. Receptors, located in skeletal muscle, are called _____

_____. They are sensitive to changes in

_____.

This type of reflex might therefore be called a _____ reflex.

d. Since sensory impulses enter the cord on the same side as motor impulses leave, the reflex is called (*ipsilateral? contralateral? intersegmental?*).

e. What structure is the effector? _____

5. Consider the reflex that would occur if you stepped on a tack and spontaneously withdrew your foot. Explain why this reflex can be classified in each of the following categories.

a. Flexor reflex

b. Withdrawal reflex

c. Ipsilateral reflex

d. Polysynaptic reflex

e. Intersegmental reflex

6. Answer these questions about crossed extensor reflexes.

a. Are they ipsilateral? (*Yes? No?*)

b. May they be intersegmental? (*Yes? No?*)

c. Where are they found? (Give an example.)

7. Reciprocal inhibition is vital in coordinating body movements. Explain how reciprocal inhibition enables antagonistic muscles to work in synchrony, so that extensors relax while flexors contract.

(pages 398–407) **D. Spinal nerves**

9 1. Complete this exercise about spinal nerves.

   a. Spinal nerves are attached by two roots. The posterior root is (*sensory? motor? mixed?*), the anterior root is _____, whereas the spinal nerve is _____ .

   b. Spinal nerves branch when they leave the intervertebral foramen. These branches are called _____ . Since they are extensions of spinal nerves, rami are (*sensory? motor? mixed?*).

   c. Which ramus is larger? (*ventral? dorsal?*) What areas does it supply?

   d. What area does the dorsal ramus innervate?

   e. Name two other branches (rami) and state their functions.

   f. There are ____ pairs of spinal nerves. Write the number of pairs in each region.

      _____ cervical

      _____ thoracic

      _____ lumbar

      _____ sacral

      _____ coccygeal

   g. The (*ventral? dorsal?*) rami of most spinal nerves form networks called

      _____ .

2. Summarize the names, origins, and general distribution of the plexuses by completing this table.

Name of Plexus	Spinal Nerves Forming Plexus	General Area Supplied by Plexus
a.		Neck, scalp, shoulder, diaphragm
b. Brachial		
c.	L1–L4	
d.		Posterior of thigh, most of leg and foot

3. Match the plexus names with descriptions.

_____ a. Provides the entire nerve supply for the arm

_____ b. Contains origin of phrenic nerve

_____ c. Formed by ventral rami of L4–S3.

_____ d. Forms median, radial, and axillary nerves

_____ e. Injury prevents normal flexing of thigh (as in touching toes)

_____ f. Forms roots, trunks, divisions, cords, and then nerves

_____ g. Not a plexus at all, but rather segmentally arranged nerves.

_____ h. Supplies nerves to scalp, neck, and part of shoulder and chest

_____ i. Supplies the posterior of thigh and most of leg and foot

B. Brachial
C. Cervical
I. Intercostal
L. Lumbar
S. Sacral

4. Refer to Figure 15-11 in the text. Note that some nerves formed from the brachial plexus contain fibers from each level between C5 and T1. Others contain fibers originating from only a single segment of the cord. Try to follow the branching pattern that forms each of the nerves listed below, and identify which spinal cord segments contribute fibers to these nerves. Verify your findings by checking origins in Exhibit 15-3.

206

a. Ulnar nerve

b. Radial nerve

c. Dorsal scapular nerve

5. If the cord were completely transected (severed) just below the C7 spinal nerves, how would the functions listed below be affected? (Remember that nerves which originate below this point would not communicate with the brain, so would lose much of their function.) Explain your reasons in each case.

11

a. Breathing via diaphragm

b. Movement and sensation of thigh and leg

c. Flexion and adduction of the arm by means of the pectoral muscle. (Hint: Notice that two nerves supply this muscle.)

d. Abduction of the arm

e. Flexion of forearm by biceps brachii and brachialis

f. Use of triceps brachii for extension of forearm

g. Use of muscles of facial expression and muscles that move jaw, tongue, eyeballs

6. Perform the actions caused by each of the muscles listed below. Remember that skeletal muscles will not contract, no matter how much you wish them to, if their nerve supply is lost. Name the nerves that must be intact for each of these muscles to function. Check your answers by referring to the exhibits listed.

a. Hamstrings (Exhibits 10-19, 15-5)

b. Dorsiflexors of foot (Exhibits 10-20, 15-5)

c. Quadriceps femoris and adductor group of thigh (Exhibits 10-19, 15-4)

d. Flexors of fingers (Exhibits 10-16, 15-3)

e. Gluteus maximus (Exhibits 10-18, 15-5)

208

7. What is a *dermatome?*

8. In general, describe the pattern of dermatomes:
   a. In the trunk

   b. In the extremities

(pages 407–410)   **E. Applications to health**

1. Contrast the terms *paraplegic* and *quadriplegic.*

12 2. Answer these questions about nerve regeneration.

   a. In order for damaged neurons to be repaired, they must have an intact
      cell body and also a _____.
   b. Can axons in the CNS regenerate? Explain.

   c. When a nerve fiber (axon or dendrite) is injured, the changes that follow
      in the cell body are called (*axon reaction? Wallerian degeneration?*).
      Those that occur in the portion of the fiber distal to the injury are known

      as _____.

3. Explain how peripheral nerves regenerate by describing each of these
   events.
   a. Axon reaction, chromatolysis

16

The Brain and the Cranial Nerves

In the previous chapter you began a study of major nervous structures with a consideration of the spinal cord and its nerves. In this chapter you will learn about the brain, the structure involved in higher levels of integration and control. You will first identify principal parts of the brain (Objective 1) and the factors that are protective to the brain (2–4). Then you will consider in more detail the major parts of the brain (5–13) and the cranial nerves (14–16). Finally you will learn about disorders and medical terminology associated with the central nervous system (17–18).

Topics Summary

A. Brain: principal parts, protection and coverings, cerebrospinal fluid, blood supply

B. Brain: brain stem

C. Brain: cerebrum, cerebellum

D. Cranial nerves

E. Applications to health

Objectives

1. Identify the principal areas of the brain.
2. Describe the location of the cranial meninges.
3. Explain the formation and circulation of cerebrospinal fluid.
4. Describe the blood supply to the brain and the concept of the blood-brain barrier.
5. Compare the components of the brain stem with regard to structure and function.
6. Identify the structure and functions of the diencephalon.
7. Identify the structural features of the cerebrum.
8. Define the lobes, tracts, and basal ganglia of the cerebrum.
9. Describe the structure and functions of the limbic system.
10. Compare the motor, sensory, and association areas of the cerebrum.
11. Describe the principle of the electroencephalograph and its significance in the diagnosis of certain disorders.
12. Explain brain lateralization and the split-brain concept.
13. Describe the anatomical characteristics and functions of the cerebellum.
14. Define a cranial nerve.
15. Identify the 12 pairs of cranial nerves by name, number, type, location, and function.
16. Explain the effects of injury on cranial nerves.
17. List the clinical symptoms of these disorders of the nervous system: poliomyelitis, syphilis, cerebral palsy, Parkinsonism, epilepsy, multiple sclerosis, cerebrovascular accidents, dyslexia, Tay-Sachs disease, headache, and trigeminal neuralgia.
18. Define key medical terms associated with the central nervous system.

(pages 414–419) **A.** Brain: principal parts, protection and coverings, cerebrospinal fluid, blood supply

1. List the four principal parts of the brain.

1 2. Identify brain structures on Figure LG 16-1. Then complete this exercise.

a. Structures 1–3 are parts of the _____ .

b. Structures 4 and 5 together form the _____ .

c. Structure 6 is the largest part of the brain, the _____ .

d. The second largest part of the brain is structure 7, the _____ .

2 3. Name three ways in which the brain is protected.

4. Review the layers of the meninges by listing them here.

5. In what spaces is cerebrospinal fluid (CSF) located?

3 6. Circle the correct answers about CSF.

a. How much CSF is found in the entire nervous system?
A. 16 oz (1 pint)   B. 8 oz (1 cup)   C. 4 oz (½ cup)
D. 1 oz (2 tablespoons)

b. The color of CSF is:
A. Yellow   B. Clear, colorless   C. Red   D. Green

c. Choose the function(s) of CSF:
A. Serves as a shock absorber for brain and cord
B. Contains red blood cells
C. Contains white blood cells
D. Contains nutrients

d. Which statement(s) describe its formation and pathway?
A. It is formed by diffusion of substances from blood.
B. It is formed by active transport.
C. It is formed by filtration.
D. It is formed from blood in capillaries called choroid plexuses.
E. It is formed in all four ventricles.

e. Which statement(s) describe the pathway of CSF?
A. It circulates around the brain, but not the cord.
B. It flows below the end of the spinal cord.
C. It bathes the brain by flowing through the epidural space.
D. It passes via projections (villi) of the arachnoid into blood vessels
(venous sinuses) surrounding the brain.
E. It is formed initially from blood, and finally flows back to blood.

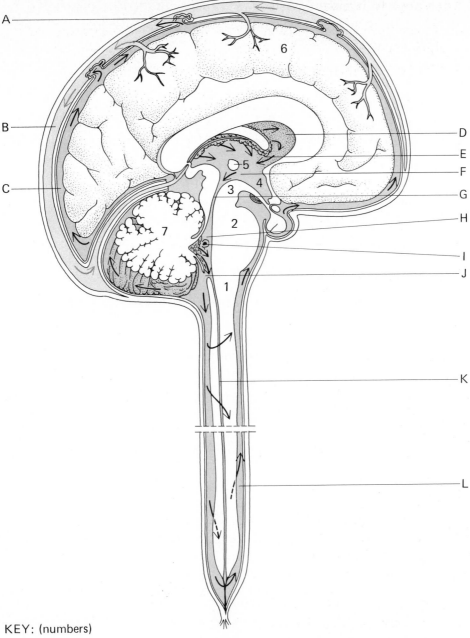

KEY: (numbers)

1. Medulla oblongata
2. Pons varolii
3. Midbrain
4. Hypothalamus
5. Thalamus
6. Cerebrum
7. Cerebellum

KEY: (letters)

A. Arachnoid villus
B. Cranial venous sinus
C. Subarachnoid space of brain
D. Lateral venticle
E. Interventricular foramen (of Monro)
F. Third ventricle
G. Cerebral aqueduct (of Sylvius)
H. Fourth ventricle
I. Lateral aperture (Foramen of Luschka)
J. Median aperture (Foramen of Magendie)
K. Central canal
L. Subarachnoid space of spinal cord

Figure LG 16-1 Brain and meninges seen in sagittal section. Parts of the brain are numbered; refer to question 1 . Key letters indicate structures in pathway of cerebrospinal fluid; refer to question 4 .

7. To check your understanding of the pathway of CSF, list in order key letters on Figure LG16-1, indicating the structures through which CSF passes. Start at the site of formation.

4

8. Contrast *internal hydrocephalus* with *external hydrocephalus*.

9. What important substances are carried by blood to the brain?

10. How do brain blood capillaries differ from other blood capillaries?

What is the significance of this difference?

(pages 419–429) **B. Brain: brain stem**

1. Briefly describe each of the functions of the medulla. Use these key words in your descriptions.
   a. Conduction (tracts)

   b. Decussation

   c. Nucleus gracilis and nucleus cuneatus

d. Reticular formation

e. Vital centers

f. Nonvital centers

g. Cranial nerves

2. List six functions of the pons.

3. Describe the functions of the midbrain.

4. Describe the thalamus according to:
   a. Structure

b. Location (see Figure LG 16-1)

c. Main functions

5. The name *hypothalamus* indicates that this structure lies (*above? below?*) the thalamus.
6. The hypothalamus is vital to the control of homeostasis. Explain the ways in which it accomplishes this.

7. Check your understanding of these parts of the brain stem and diencephalon by matching them with descriptions given below.

5

_____ a. It is the main regulator of visceral activities since it acts as a liaison between cerebral cortex and autonomic nerves that control viscera.

_____ b. It is the site of the red nucleus, the origin of rubrospinal tracts concerned with muscle tone and posture.

_____ c. Cranial nerves III–IV attach to this brain part.

_____ d. Cranial nerves V–VIII attach to this brain part.

H. Hypothalamus
Med. Medulla
Mid. Midbrain
P. Pons
T. Thalamus

_____ e. Cranial nerves IX–XII attach to this brain part.

_____ f. Feelings of hunger, fullness, and thirst stimulate centers here so that you can respond accordingly.

_____ g. All sensations except smell are relayed through here.

_____ h. Regulation of heart, blood pressure, and respiration occurs here.

_____ i. Constitutes four-fifths of the diencephalon.

_____ j. It lies under the third ventricle, forming its floor.

_____ k. It forms most of side walls of the third ventricle.

_____ l. Tumor in this region could compress cerebral aqueduct and cause internal hydrocephalus.

_____ m. It releases chemicals called regulating factors that control hormones.

C. Brain: cerebrum, cerebellum

(pages 429–439)

1. Complete this exercise about cerebral structure.

[6]

a. The outer layer of the cerebrum is called _____. It is composed of (*white? gray?*) matter. This means that it contains mainly (*cell bodies? neurons?*).

b. In the right margin, draw a line the same length as the thickness of the cerebral cortex. Use a metric ruler. Note how thin the cortex is.

c. The surface of the cerebrum looks much like a view of tightly packed mountains or ridges, called _____. The parts where the cerebral cortex dips down into valleys are called _____

(deep valleys) or _____ (shallow valleys).

d. The cerebrum is divided into two halves called _____. Connecting the two hemispheres is a band of (*white? gray?*) matter called

the _____. Notice this structure in Figure LG 16-1 and in Figure 16-6 of your text.

e. The falx cerebri is composed of (*nerve fibers? dura mater?*). Where is it located?

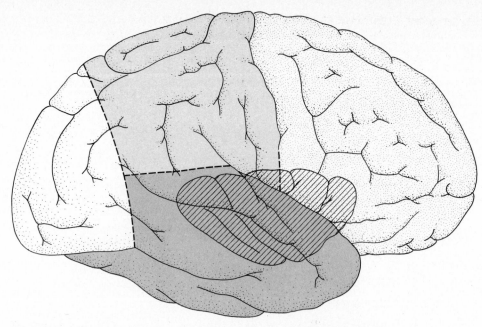

Figure LG 16-2 Right lateral view of lobes and fissures of the cerebrum. Color and label as directed.

2. Label the following cerebral structures on the lateral view in Figure LG 16-2: *frontal lobe, parietal lobe, occipital lobe, temporal lobe, central sulcus, lateral sulcus, precentral gyrus,* and *postcentral gyrus.*

7 | 3. Match the three types of white matter fibers with these descriptions.

_____ a. The corpus callosum contains these fibers and connects the two cerebral hemispheres.

A. Association
C. Commissural
P. Projection

_____ b. Sensory and motor fibers passing between cerebrum and other parts of the CNS are this type of fiber.

_____ c. These fibers transmit impulses among different areas of the same hemisphere.

8 | 4. Look at Figures 16-6 and 16-10 in the text and identify the parts of the *corpus striatum* by completing this description.

a. A structure with a head and a tail (cauda) is the _____ nucleus.

b. The two parts of the lentiform nucleus are the _____ and _____ .

c. Together these three areas are called the _____ . They are islands of (*gray? white?*) matter embedded deep in the cerebrum.

d. Separating the caudate from the lentiform nucleus is a band of white fibers called the _____ . It consists of important sensory and motor tracts. All of these structures together present a striped appearance (gray caudate, white internal capsule, gray lentiform nucleus) called the _____ .

5. In general, what are the functions of the basal ganglia?

What are the results of damage to it?

6. Describe the limbic system.

   a. Name its components.

   b. Tell how it helps to control behavior.

7. Use this list of functional areas of the brain in two ways. First, write a sentence next to each name describing the functions of that area. Second, color (as indicated below) and label those areas on Figure LG 16-2.

   a. General sensory (somesthetic) area, blue

   b. Somesthetic association area, light blue

   c. Primary visual area, green

d. Visual association area, light green

e. Primary auditory area, black

f. Auditory association area, gray

g. Primary gustatory area, purple

h. Primary motor area, red

i. Premotor area, pink

j. Frontal eyefield area, brown

k. Broca's area, yellow

8. What is the function of the gnostic area?

9. List the sequence of events that occurs when you listen to a person ask a question and then you respond to it orally. (Notice how many parts of the cerebrum are involved.)

10. Contrast:
   a. Aphasia/agraphia

   b. Word blindness/word deafness

11. What do the letters *EEG* stand for? _____ Briefly describe an *EEG*.

   9

12. Define *brain lateralization*.

Explain how brain lateralization relates to each of the following.
a. Anatomical differences between hemispheres in a right-handed person

b. Language ability

c. Ability to sing

d. Emotions

13. Describe the cerebellum according to:
a. Location

b. Structure (Identify the *vermis* and *hemispheres, arbor vitae* and *peduncles.*)

c. Functions (Explain how it relates to the cerebrum in controlling coordinated muscle activity, posture, and muscle tone.)

D. Cranial nerves

1. Answer these questions about cranial nerves.   ☐10

   a. There are _____ pairs of cranial nerves. They are all attached to the
      _____ ; they leave the _____ via
      foramina.
   b. They are numbered by Roman numerals in the order that they leave the
      cranium. Which is most anterior? (*I? XII?*) Which is most posterior (*I?
      XII?*)
   c. All spinal nerves are (*purely sensory? purely motor? mixed?*). Are all cra-
      nial nerves mixed? (*Yes? No?*)

2. Pathways involving cranial nerves are quite similar to those of spinal
   nerves, but are shorter since they are closer to the brain. Describe the typical sensory and motor pathways of cranial nerves in this exercise.   ☐11

   a. Cranial nerves with somatic motor fibers have relatively simple path-
      ways, as shown in Figure LG 16-3a. Like spinal nerve pathways, they
      consist of just one neuron between the CNS (brain stem in this case) and
      effector. (Autonomic pathways involve two neurons; these will be dis-
      cussed in Chapter 17.) Motor cell bodies that lie in (*ganglia? nuclei?*) in
      the brain stem send axons via cranial nerves to effectors. (These cell
      bodies could be stimulated by a variety of neurons in other parts of the
      brain.)
   b. The typical sensory pathway is shown in Figure LG 16-3b. It involves
      (*axons? dendrites?*) that extend from receptor organs to cell bodies in
      ganglia situated just (*inside? outside?*) the CNS. Axons from these cell
      bodies proceed into the CNS (brain stem) and the cranial nerves termi-
      nate in (*ganglia? nuclei?*) there. Then neurons relay impulses via tracts to

      the _____ , which passes messages to the cerebral

      _____ . The entire conduction pathway from recep-

      tor to cerebral cortex involves at least _____ neurons.

3. Pathways for smell and sight are more complicated. The olfactory pathway
   is shown on Figure LG 16-3c. Describe it in this exercise.   ☐12

   a. Cell bodies of olfactory nerves are located in the _____ .
   b. How many neurons are in the olfactory pathway from the nose to the ol-

      factory area of the cerebral cortex? _____
   c. Which fibers are longer? (*Olfactory nerves? Olfactory tract?*)

4. Answer these questions about the visual pathway, shown on Figure LG
   16-3d.   ☐13

   a. There are _____ layers of neurons in the retina of the eye.

   b. Axons of the third layer, namely the _____ cells,

      form the second cranial nerve, known as the _____
      nerve.
   c. The optic nerve exits from the eye and joins its partner from the other

      eye at the optic _____ . Here some fibers cross to the
      opposite side; all optic fibers proximal to this point are located in the

      optic _____ .

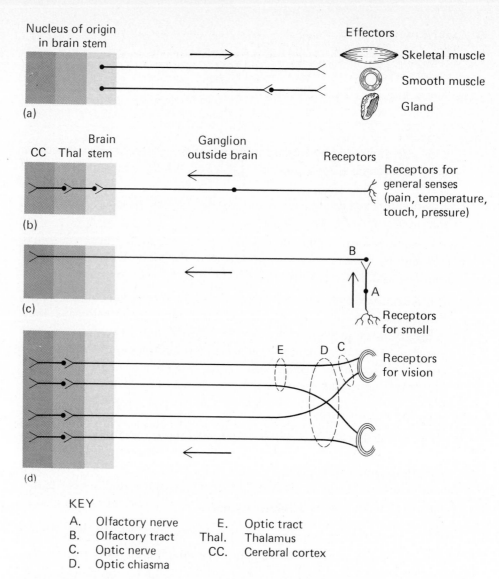

KEY
A. Olfactory nerve   E.   Optic tract
B. Olfactory tract   Thal. Thalamus
C. Optic nerve   CC.   Cerebral cortex
D. Optic chiasma

Figure LG 16-3 Conduction pathways of cranial nerves. (a) Typical motor pathway. (b) Typical sensory pathway. (c) Olfactory pathway. (d) Visual pathway.

d. Most optic tract fibers terminate in the _____ where

impulses are relayed to the visual areas of the _____.

e. Some optic tract fibers do not go to the thalamus, but rather lead to the

_____ , where they cause reflex action that causes

eye _____ in response to visual stimuli.

5. Complete the table about cranial nerves.

Number	Name	Functions
a.	Olfactory	

Number	Name	Functions
b.		Vision (not pain or temperature of the eye)
c. III		
d.	Trochlear	
e. V		
f.		Stimulates lateral rectus muscle to abduct eye; proprioception of the muscle
g.	Facial	
h.		Hearing; equilibrium
i. IX		
j.	Vagus	
k. XI		
l.		Supplies muscles of the tongue with motor and sensory fibers

6. Check your understanding of cranial nerves by completing this exercise. Write the name of the correct cranial nerve following the related description.

[14]

   a. Differs from all other cranial nerves in that it originates from the brain stem and from the spinal cord: _____

   b. Eighth cranial nerve (VIII): _____

   c. Is widely distributed into neck, thorax, and abdomen: _____

   d. Senses toothache, pain under a contact lens, wind on the face: _____

   e. The largest cranial nerve; has three parts (ophthalmic, maxillary, and mandibular): _____

   f. Controls contraction of muscle of the iris, causing constriction of pupil: _____

   g. Innervates muscles of facial expression: _____

   h. Two nerves that contain taste fibers and autonomic fibers to salivary glands: _____ , _____

   i. Three purely sensory cranial nerves: _____ , _____ , _____

7. Write the number of the cranial nerve related to each of the following disorders.

[15]

   _____ a. Bell's palsy

   _____ b. Inability to shrug shoulders or turn head

   _____ c. Anosmia

   _____ d. Strabismus and squinting

   _____ e. Blindness

   _____ f. Tic douloureux

   _____ g. Paralysis of vocal cords; loss of sensation of many organs

   _____ h. Vertigo and nystagmus

(pages 444–450)   **E. Applications to health**

1. For each of the following disorders, give a brief description. Name the cause(s) of the disorder, if they are known; age group or population most afflicted; parts of the nervous system affected; symptoms; and treatment. If the disorder occurs in a progressive manner, describe the stages. Also answer extra questions listed for some disorders.

   a. Poliomyelitis. How does the meaning of the term "poliomyelitis" indicate its effects?

b. Syphilis, neurosyphilis, tabes dorsalis

c. Cerebral palsy

d. Parkinsonism

e. Multiple sclerosis. How does the name give a clue to the nature of this disorder?

f. Epilepsy. Contrast types of seizures: *grand mal, petit mal, psychomotor, idiopathic.*

g. Cerebrovascular accident (cerebral apoplexy). Define and explain effects of *emboli, atherosclerosis,* and *arteriosclerosis.*

h. Headaches

16 2. Match the name of the disorder with the related description.

Cerebral palsy   Multiple sclerosis
Cerebrovascular accident   Neuralgia
  (CVA)   Neuritis
Dyslexia   Parkinsonism
Epilepsy   Poliomyelitis
Encephalitis   Tabes dorsalis
Migraine   Tay-Sachs disease

a. Degeneration of myelin sheath to form hard plaques in many (multiple) regions causing loss of motor and sensory function: _____

b. Degeneration of dopamine-releasing neurons in basal ganglia; characterized by tremor or rigidity of muscles: _____

c. A type of throbbing headache: _____

d. Most common brain disorder; also called stroke: _____

e. Characterized by difficulty in handling words and symbols, for example, reversal of letters (*b* for *d*); cause unknown: _____

f. Inflammation of the brain: _____

g. Inflammation along a nerve: _____

h. Inherited disease due to excessive lipids (gangliosides) in brain cells; affects infants: _____

i. Affects sensations, but not movement; a form of syphilis: _____

j. Of viral origin; often affects only the respiratory system; when nervous system involved, affects movement, but not sensation: _____

17
The Autonomic Nervous System

You have studied the structures and functions of the somatic nervous system in the preceding three chapters. Now you will consider control of the viscera by means of the autonomic nervous system (ANS). You will identify structural components and pathways of this system and contrast the two great divisions of the ANS (Objectives 1–4). You will consider the involvement of the hypothalamus with the ANS (5). You will also discuss relationships between biofeedback and meditation and the ANS (6–7).

Topics Summary

A. Somatic efferent and autonomic nervous systems

B. Structure

C. Physiology

D. Visceral autonomic reflexes

E. Control by higher centers

Objectives

1. Compare the structural and functional differences between the somatic efferent and autonomic portions of the nervous system.

2. Identify the structural features of the autonomic nervous system.

3. Compare the sympathetic and parasympathetic divisions of the autonomic nervous system in terms of structure, physiology, and chemical transmitters released.

4. Describe a visceral autonomic reflex and its components.

5. Explain the role of the hypothalamus and its relationship to the sympathetic and parasympathetic division.

6. Explain the relationship between biofeedback and the autonomic nervous system.

7. Describe the relationship between meditation and the autonomic nervous system.

Learning Activities

(page 456) **A. Somatic efferent and autonomic nervous systems**

| 1 | 1. Give examples of structures innervated by each type of nerve.

  a. Somatic afferent

  b. Visceral afferent

  c. Somatic efferent

  d. Visceral efferent

| 2 | 2. Which of the above types of nerve are autonomic?

3. Why is the *autonomic nervous system* so named? Is the autonomic nervous system (ANS) entirely independent of higher control centers? Explain.

4. In general, what are the structural components of the ANS?

5. Name the two divisions of the ANS.

6. Many visceral organs have *dual innervation* by the autonomic system.
   a. What does dual innervation mean?

   b. How do the sympathetic and parasympathetic divisions work in harmony to control viscera?

B. Structure (pages 456–462)

1. Complete this exercise describing differences between visceral efferent and somatic efferent pathways.

   ⬛ 3

   a. Somatic pathways begin in the (*anterior? lateral?*) gray horns of the cord, whereas visceral pathways begin in the _____ gray horns of the cord or in nuclei within the _____ .

   b. Somatic pathways begin at (*all? only certain?*) levels of the cord, whereas visceral routes begin at _____ levels of the cord.

   c. Between the spinal cord and effector, somatic pathways include ———— neuron(s), whereas visceral pathways require ———— neuron(s), known as the pre-_____ and post-_____ neurons.

2. Contrast according to location and myelination:
   a. Preganglionic neurons

   b. Postganglionic neurons

3. Another name for the sympathetic division is _____;
an alternate name for the parasympathetic system is _____
_____. These names are based on locations of (*preganglionic? post-ganglionic?*) cell bodies.

4. Contrast these two types of ganglia.

	Posterior Root Ganglia	Autonomic Ganglia
a. Location of cell bodies of (*afferent? efferent?*) neurons		
b. Location of cell bodies of neurons in (*somatic? visceral? both?*) pathways		
c. Synapsing (*does? does not?*) occur here		

5. Complete the table contrasting types of autonomic ganglia.

	Sympathetic Trunk	Prevertebral	Terminal
a. Sympathetic or parasympathetic			
b. Alternate name	Paravertebral		
c. General location			Close to or in walls of effectors

6. Identify descriptions that fit the kinds of ganglia listed below. Answers may be used more than once; some questions require more than one answer.

_____ a. Located near first rib

_____ b. Located in solar plexus

_____ c. Site of synapsing of neurons in lesser splanchnic nerves

_____ d. At level of C3 vertebra; supply the head with postganglionic sympathetic fibers

_____ e. Supply heart with nerve fibers

_____ f. Types of prevertebral ganglia

_____ g. Sympathetic ganglia

C. Celiac
IC. Inferior cervical
IM. Inferior mesenteric
MC. Middle cervical
SC. Superior cervical
SM. Superior mesenteric

7. Describe the pathway common to all sympathetic preganglionic neurons by listing the structures in correct sequence. (See Figure 17-3 in your text.) | 6 |

 ____ ____ ____ ____

    A. Anterior root of spinal nerve
    B. Sympathetic trunk ganglion
    C. Lateral gray of spinal cord (between T1 and L2)
    D. White ramus communicantes

8. Now consider the possible routes sympathetic preganglionic neurons may take once they are in the sympathetic trunk ganglion. (It may be helpful to refer to Figure 17-3 in your text.) | 7 |

   a. What is the shortest possible path they can take to reach a postganglionic neuron cell body?

   b. They may ascend and/or descend the sympathetic chain to reach trunk ganglia at other levels. This is important since sympathetic preganglionic cell bodies are limited in location to _____ levels of the cord, yet the sympathetic trunk extends from _____

    _____ to _____ levels of the vertebral column. Sympathetic fibers must be able to reach these distant areas to provide the entire body with sympathetic innervation.

   c. Viscera in skin and extremities (such as sweat glands, blood vessels, hair muscles) receive sympathetic innervation through the following route. Preganglionic neurons synapse with postganglionic cell bodies in sympathetic trunk ganglia (at the level of entry or after ascending or descending). Postganglionic fibers then pass through _____

    which connect to _____ .

   d. Some preganglionic fibers do not synapse as described in (a) or (b), but pass on through trunk ganglia without synapsing there. They course

    through _____ nerves to _____ ganglia. Here they synapse with postganglionic neurons whose axons

    form _____ en route to viscera.

   e. Note that prevertebral ganglia are located only in the _____

    _____ . In the neck, thorax, and pelvis the only sympathetic ganglia are those of the trunk (see Figure 17-2 in the text). As a result, cardiac nerves, for example, contain only (*preganglionic? postganglionic?*) fibers.

   f. A given sympathetic preganglionic neuron is likely to have (*few? many?*) branches; these may take any of the paths described. Once a *branch* synapses, it (*can? cannot?*) synapse again, since any autonomic pathway

    consists of just _____ neurons (preganglionic and postganglionic).

9. Test your understanding of these routes by completing the sympathetic pathway shown in Figure LG 17-1. A preganglionic neuron located at level T5 of the cord has its axon drawn as far as the white ramus. Finish this pathway by showing how the axon may branch and synapse to eventually

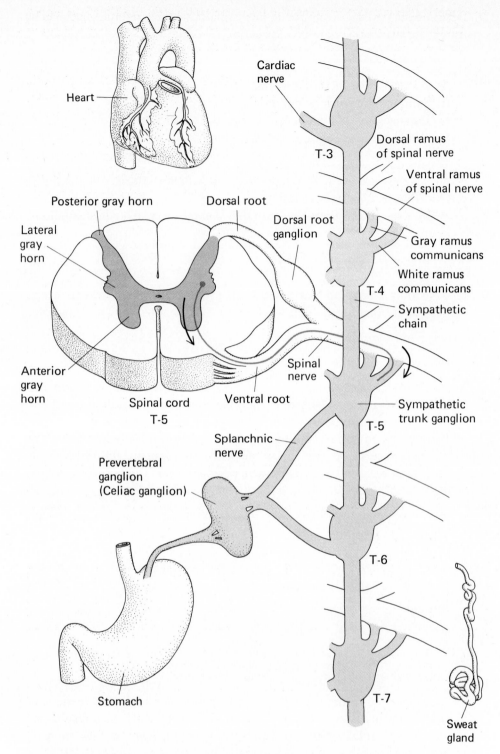

Figure LG 17-1 A typical sympathetic pathway beginning at level T5 of the cord. Complete pathway to the heart, a sweat gland in the skin of the upper thoracic region, and the stomach.

innervate these three organs: the heart, a sweat gland in skin of the upper thoracic region, and the stomach. Draw the preganglionic fibers in solid lines and the postganglionic fibers in broken lines.

10. Now that you are somewhat familiar with the rather complex sympathetic pathways, you will notice that parasympathetic pathways are much simpler. This is due to the fact that there is no parasympathetic chain of ganglia; the only parasympathetic ganglia are _____, which are in or close to the organs innervated. Also parasympathetic pre-ganglionic fibers synapse with (*more? fewer?*) postganglionic neurons. What is the significance of the structural simplicity of the parasympathetic system?

8

11. Complete the table about the cranial portion of the parasympathetic system.

Cranial Nerve		Name of Terminal Ganglion	Structures Innervated
Number	Name		
a.			Iris and ciliary muscle of eye .
b.	Facial		
c.		Otic ganglion	
d. X			

12. Describe the pathway of neurons in the sacral portion of the parasympathetic system. Tell what organs they innervate.

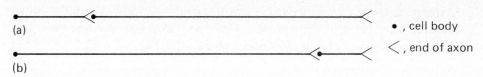

(a)

(b)

•, cell body

<, end of axon

Figure LG 17-2 Comparison of preganglionic and postganglionic neurons. (a) Sympathetic; (b) Parasympathetic. Complete as directed.

(pages 462–463) **C. Physiology**

1. Refer to Figure LG 17-2 and do this exercise.
   a. In general, which system has long preganglionic fibers and short postganglionic fibers? (*Sympathetic? Parasympathetic?*) Identify the pathways accordingly.
   b. Write *C* next to ends of axons that are *cholinergic*.
   c. Sympathetic postganglionic fibers to certain effectors are cholinergic. In the list below identify effectors as cholinergic (C) or adrenergic-stimulated (A).

   _____ Sweat glands   _____ Blood vessels in skeletal muscles

   _____ Blood vessels in skin   _____ Blood vessels in genitalia

   _____ Smooth muscle of digestive organs

2. Which of the two transmitter substances has longer lasting and more widespread effects? (*Acetylcholine? Noradrenaline?*)
3. Use arrows to show whether parasympathetic (P) or sympathetic (S) fibers stimulate (↑) or inhibit (↓) each of the following activities. Use a dash (—) to indicate that there is no parasympathetic innervation. The first one is done for you.

   9

   a. P __↓__   S __↑__   Dilation of pupil

   b. P _____   S _____   Heart rate and blood flow to coronary (heart muscle) blood vessels

   c. P _____   S _____   Constriction of skin blood vessels

   d. P _____   S _____   Salivation and digestive organ contractions

   e. P _____   S _____   Sweating

   f. P _____   S _____   Dilation of bronchioles for easier breathing

   g. P _____   S _____   Coping with stress, "fight or flight"

   h. P _____   S _____   Conservation of energy, rest

   i. P _____   S _____   Release of epinephrine (adrenaline)

(pages 464) **D. Visceral autonomic reflexes**

1. On Figure LG 17-1 draw an afferent neuron (in contrasting color) to show how the sympathetic neuron could be stimulated.
2. Autonomic neurons can be stimulated by (*somatic afferents? visceral afferents? either somatic or visceral afferents?*).

   10

3. Arrange in correct sequence structures in the pathway for a painful stimulus at your fingertip to cause a visceral (autonomic) reflex, such as sweating. Write the letters of the structures in order on the lines provided.

11

___ ___ ___ ___ ___ ___ ___ ___ ___ ___

  A. Association neuron in spinal cord
  B. Nerve fiber in gray ramus
  C. Pain receptor in skin
  D. Cell body of postganglionic neuron in trunk ganglion
  E. Cell body of preganglionic neuron in lateral gray of cord
  F. Nerve fiber in anterior root of spinal nerve
  G. Nerve fiber in white ramus
  H. Fiber in spinal nerve in brachial plexus
  I. Sweat gland
  J. Sensory neuron

4. Most visceral sensations (*do? do not?*) reach the cerebral cortex, so most visceral sensations are at (*conscious? subconscious?*) levels. Hunger and thirst are exceptions.

E. Control by higher centers

(pages 464–465)

1. Explain the role of each of these structures in control of the autonomic system.
   a. Thalamus

   b. Hypothalamus (Which part controls sympathetic nerves? Which controls parasympathetic?)

   c. Cerebral cortex (When is it most involved in ANS control? In stress or nonstress situations?)

2. Define *biofeedback*.

3. Explain how biofeedback is used to help in each of the following cases.
   a. When heart rate is too fast

   b. During migraine headache

   c. In childbirth

4. Explain how biofeedback and yoga may demonstrate whether or not the autonomic nervous system is truly autonomous.

Answers to Numbered Questions in Learning Activities

1. There are many examples possible; here are a few. (a) Skin exteroceptors (for pain, heat, cold, touch) and proprioceptors in muscles, tendons, and joints. (b) Visceroceptors, as in heart, digestive organs, walls of blood vessels. (c) Skeletal muscles, such as gastrocnemius. (d) Smooth muscle (digestive organs, urinary bladder, walls of blood vessels, hair muscles), cardiac muscle (heart), and glands (sweat, salivary, other digestive glands).

2. Visceral efferent.

3. (a) Anterior, lateral, brain stem. (b) All, only certain (sympathetic: T1 to L2; parasympathetic: S2 to S4 plus brain stem). (c) 1, 2, ganglionic, ganglionic.

4. (a) Afferent, efferent. (b) Both, visceral. (c) Does not, does.

5. (a) IC. (b) C. (c) SM. (d) SC. (e) IC, MC, SC. (f) C, IM, SM. (g) All.

18

Sensory Structures

In the previous chapter you learned about the structure and function of the many individual parts of the brain. In this chapter you will see how these parts work together with the spinal cord, nerves, and receptors to enable the body to receive sensory information (Objectives 1–7). You will learn about special receptors and pathways that make you more aware of your external environment, those of the special senses: smell (Objective 8), taste (9), vision (10–12), hearing (13–14), and equilibrium (15). Then you will study disorders and key medical terms associated with the special senses (16–17). .

Topics Summary

A. Sensations
B. General senses
C. Sensory and motor pathways
D. Olfactory sensations
E. Gustatory sensations
F. Visual sensations
G. Auditory sensations and equilibrium
H. Applications to health; key medical terms

Objectives

1. Define a sensation and list the four prerequisites for its transmission.
2. Compare the location and function of exteroceptors, visceroceptors, and proprioceptors.
3. Describe the distribution of cutaneous receptors.
4. List the location and function of the receptors for tactile sensation (touch, pressure, vibration), thermoreceptive sensations (heat and cold), pain, and proprioception.
5. Distinguish among somatic, visceral, referred, and phantom pain.
6. Describe how acupuncture is believed to relieve pain.
7. Identify the locations and components of. the principal sensory and motor pathways.
8. Locate the receptors for olfaction and describe the neural pathway for smell.
9. Identify the gustatory receptors and describe the neural pathway for taste.
10. Describe the structure of the accessory visual organs.
11. List the structural divisions of the eye.
12. Describe the afferent pathway of light impulses to the brain.
13. Define the anatomical subdivisions of the ear.
14. List the principal events in hearing.
15. Identify the receptor organs for static and dynamic equilibrium.
16. Contrast the causes and symptoms of cataract, glaucoma, conjunctivitis, trachoma, labyrinthine disease, Ménière's disease, impacted cerumen, otitis media, and motion sickness.
17. Define key medical terms associated with the sense organs.

Learning Activities

(page 469) **A.** Sensations

1. Think of the ways in which your senses help you make homeostatic adjustments during the day. Suppose that you suddenly lost your sight and your hearing; then your ability to sense smell and taste disappeared; then you lost perception of touch, hot and cold, and pain sensations. Consider how you would survive with such sensory deprivation.

2. Define:
   a. Sensation

   b. Perception

3. List and describe four prerequisites for perception of a sensation.

4. Contrast *generator (receptor) potential* with *action potential.*

5. Answer the following questions about receptors.
   a. Define *receptor.*

   b. Give two examples of simple receptors and two of complex receptors.

c. What do all receptors have in common?

6. Identify the class of receptor that fits each description given.   2

    _____ a. Receptors in this class are located   E. Exteroceptor
              near the body surface.   P. Proprioceptor
                                                       V. Visceroceptor
    _____ b. These receptors inform you that
              you are hungry or thirsty.

    _____ c. Your ears, eyes, and receptors in
              your skin for pain, touch, hot,
              and cold are of this type.

    _____ d. With your eyes closed, you can
              tell your exact body position,
              thanks to these receptors.

    _____ e. Receptors in your blood vessels
              inform your brain of adjustments
              in heart rate and blood pressure
              needed to maintain homeostasis.

7. Identify the types of sensations listed below as *general* (G) or *special* (S).

    _____ Touch   _____ Taste
    _____ Hearing   _____ Heat
    _____ Cold   _____ Sight
    _____ Pain

B. General senses   (pages 469–473)

1. Name the cutaneous senses.

2. Explain how the *two-point discrimination test* determines the density of cutaneous receptors in different parts of the body.

3. Distinguish among the cutaneous receptors by completing the table.

Name of Receptor	Description or Sketch	Functions
a.		Two-point discrimination
b.		Touch
c. Root hair complex		
d.		Deep pressure
e.	Simple; branching ends of dendrites	

4. State examples of:
   a. Superficial somatic pain

   b. Deep somatic pain

   c. Visceral pain

5. Describe how visceral pain is referred, and explain how an understanding of referred pain is necessary for correct diagnosis.

6. To what regions does pain in each of these viscera refer? Locate these regions on your own body. Verify your answers on Figure 18-2 in the text.
   a. Gallbladder

   b. Heart

   c. Kidney

   d. Appendix ·

7. What is *phantom pain?*

8. Name five surgical methods for reducing pain.

9. Describe how *acupuncture* is used to inhibit pain. Discuss:
   a. The procedure

b. The *gate control theory* that attempts to explain how acupuncture works

10. Define *proprioception*.

3 11. Name the type of proprioceptor used in each case.

    a. Indicates changes in position of your hip and ankle joints as you do a sit-up: _____

    b. Senses tension applied to a tendon, as when you isometrically contract and relax your gluteus muscles: _____

    c. Senses the amount of stretching of your biceps and triceps brachii muscles while you flex your forearm: _____

12. Compare the levels of sensation involving the spinal cord, brain stem, thalamus, and cerebral cortex. (You will find that all of these will be involved in the sensory pathways you will study next.)

(pages 473–478)   **C. Sensory and motor pathways**

1. Complete Table LG 18-1 a–c, describing pathways controlling sensation of the left hand. Give the name of the pathway and the locations of cell bodies in each case. Be sure to indicate whether cell bodies are on the *left* or *right* side of the body. Circle the cell body in each pathway whose axons cross (decussate) to the opposite side. Refer to Figures 18-3 to 18-5 in the text to verify your answers.

2. Check your understanding of these pathways by drawing them according to the directions below. Try to draw them from memory; do not refer to the text figures or the table until you finish. Label the *first-*, *second-*, and *third-order neurons.*

    a. Use Figure LG 18-1 to show the pathway for pain and temperature in the left hand. Draw the pathway in red.

    b. Also draw on Figure LG 18-1 the pathway for fine touch, two-point discrimination, and proprioception in the left hand. Use blue for this pathway.

Table LG 18-1
Principal Sensory and Motor Pathways

Function	Name of Pathway	Location of Cell Bodies		
		First Order	**Second Order**	**Third Order**
a. Sense of pain and temperature in the *left* hand		Posterior root ganglion, *left* side	Posterior gray horn, *left* side	Thalamus, *right* side
b. Crude touch and pressure in *left* hand				
c. Two-point discrimination and proprioception in *left* hand				
			Upper Motor Neuron	**Lower Motor Neuron**
d. Movement of *left* hand	Lateral corticospinal			
e. Movement of *left* hand	Anterior corticospinal			

3. Distinguish between *crude touch* and *fine touch*.

4. Answer these questions about the spinocerebellar tracts.

    a. They are concerned with (*conscious? subconscious?*) muscle sense.

    b. They permit reflex adjustments for _____ and _____ .

    c. The left posterior spinocerebellar tract conveys impulses from muscles on (*the left side? the right side? both sides?*) of the body.

    d. The left anterior spinocerebellar tract conveys impulses from muscles on (*the left side? the right side? both sides?*) of the body.

5. Match the structures with their roles in controlling movement.

    _____ a. Controls precise, discrete movements

    _____ b. Integrates semivoluntary movements, such as walking and swimming

    _____ c. Not a control center, but assists in making movements smooth and coordinated

    B. Basal ganglia
    C. Cerebellum
    CC. Cerebral cortex

6. Show the route of impulses along the principal pyramidal pathway by listing in correct sequence the structures that comprise the pathway. Refer to Figure 18-6 in the text if you have difficulty with this activity.

___ ___ ___ ___ ___ ___ ___ ___

    A. Anterior gray horn (lower motor neuron)
    B. Midbrain and pons
    C. Effector (skeletal muscle)
    D. Internal capsule

    E. Lateral corticospinal tract
    F. Medulla, decussation site
    G. Precentral gyrus (upper motor neuron)
    H. Ventral root of spinal nerve

7. Now draw this pathway on Figure LG 18-2. Show the control of muscles in the left hand by means of neurons in the lateral corticospinal pathway. Label the *upper* and *lower motor neurons.*

8. Why are the lower motor neurons called the *final common pathway?*

Distinguish between *spasticity* and *flaccid paralysis.*

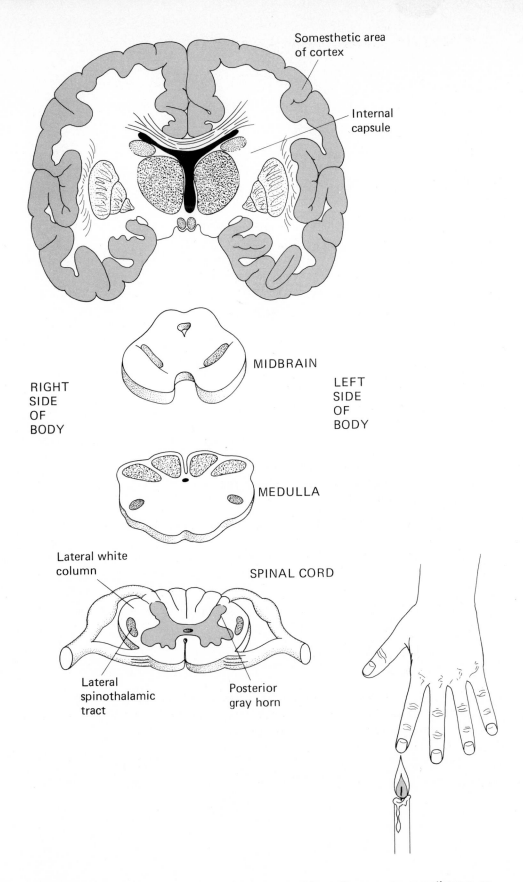

Somesthetic area
of cortex

Internal
capsule

MIDBRAIN

RIGHT
SIDE
OF
BODY

LEFT
SIDE
OF
BODY

MEDULLA

Lateral white
column

SPINAL CORD

Lateral
spinothalamic
tract

Posterior
gray horn

Figure LG 18-1 Diagram of central nervous system. Draw sensory pathways as directed.

258

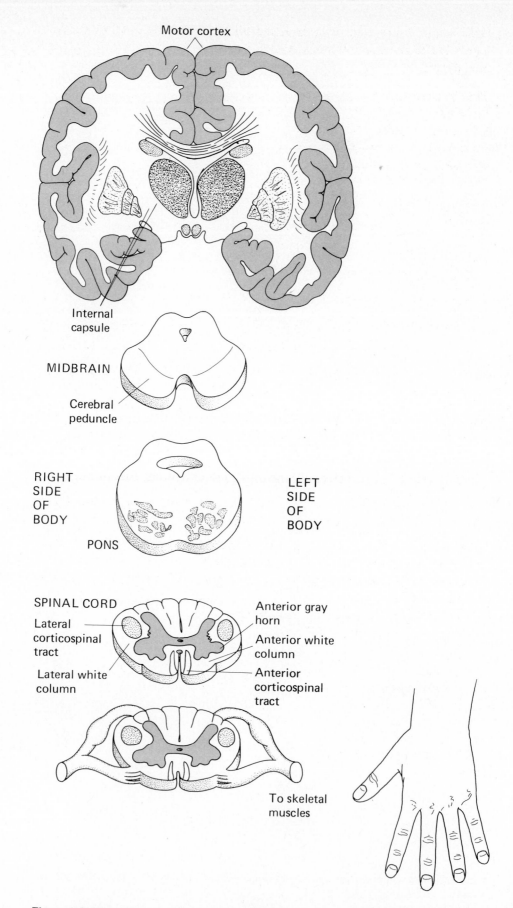

Motor cortex

Internal capsule

MIDBRAIN

Cerebral peduncle

RIGHT SIDE OF BODY

LEFT SIDE OF BODY

PONS

SPINAL CORD

Lateral corticospinal tract

Lateral white column

Anterior gray horn

Anterior white column

Anterior corticospinal tract

To skeletal muscles

Figure LG 18-2 Diagram of central nervous system. Draw motor pathway as directed.

9. How would destruction of each of the following structures affect the *left* hand? Match the effect with the related structure. Some require more than one answer.

Flaccid paralysis
Spasticity
Cannot sense pain
Cannot sense crude touch
Cannot sense discriminating touch
Little or no effect

a. *Right* thalamus: ——————————

b. Upper motor neurons on *right* side of cerebral cortex: ——————— ———————

c. CVA affecting *left* side of cerebral cortex: —————————————

d. *Left* lateral corticospinal tract at level C4 of cord: ————————————

e. *Left* lateral corticospinal tract at level T4 of cord: ———————————

f. Nucleus cuneatus on *left* side of medulla: ————————————————

g. Posterior white columns on entire *right* side of spinal cord: ————————————

h. Anterior gray horns on *left* side of spinal cord from C4 to T4, as from polio: ————————————

10. Complete this exercise about the anterior corticospinal tract.

a. Only about ——— percent of the upper motor neurons pass through anterior corticospinal tracts; ——— percent pass through the large —————————————————— corticospinal tracts.

b. Left anterior corticospinal tracts consist of axons that originated in upper motor neurons on the (*right? left?*) side of the motor cortex.

c. Contrast the two different corticospinal (pyramidal) tracts by completing Table LG 18-1, d–e.

11. What is the function of corticobulbar tracts?

Why are these tracts so named?

12. In general, the extrapyramidal tracts begin in the _____
and end in the _____ . Name three of these tracts and
their functions.

13. Summarize how the basal ganglia and cerebellum, together with association neurons, help you to carry out coordinated, precise movements, such
as those involved in a game of tennis.

Learning Activities

(pages 478–480) **D. Olfactory sensations**

1. Describe the structural arrangement of *olfactory cells* and *supporting cells* in
the nasal mucosa.

2. Review the pathway of nerve impulses for conduction of smell on Figure
LG 16-3c.

3. Discuss three theories that explain how stimuli for smell may be converted
into impulses.

4. Adaptation to odors occurs (*very rapidly? very slowly? not at all?*).

E. Gustatory sensations (pages 480–482)

1. Describe the structure and locations of taste buds.

2. Explain the relationship between *tastebuds* and *papillae*. Contrast three types of papillae by structure and location.

3. List the four taste sensations, and indicate the portions of the tongue most sensitive to each.

4. Write the number of the cranial nerve that carries sensations of taste from: $\boxed{9}$

   a. The anterior two-thirds of the tongue: _____

   b. The posterior one-third of the tongue: _____

   c. Mucosa beyond the tongue (in the throat): _____

F. Visual sensations (pages 482–489)

1. Examine your own eye structure in a mirror. With the help of Figure 18-9 in the text, identify each of the structures listed below. Then write a sentence about the function of each.
   a. Eyebrows

b. Palpebrae

c. Lateral and medial canthus

d. Lacrimal caruncle

e. Conjunctiva (palpebral and bulbar)

f. Eyelashes

2. State the locations and functions of these structures.
   a. Tarsal plate

   b. Meibomian glands

3. Discuss where tears are made and the pathway they travel. With the aid of a mirror, trace on your face the route that tears take from the lacrimal glands to the nose.

4. Tears are also known as _____ . Each day about _____ ml of this fluid is formed. Its functions are _____ . Explain why eyes may become "watery."

5. Identify the structures of the eye in Figure LG18-3. Check your answers with the key. Then color the diagram as follows: |10|
   a. Structures that are parts of the *fibrous tunic,* gray
   b. Structures that are parts of the *vascular tunic,* red
   c. Areas where *aqueous humor* is located, yellow
   d. Area where *vitreous humor* is located, blue

6. Match each of the following structures with the descriptions given. |11|

   a. "White of the eye": _____
   b. A clear structure, composed of protein layers arranged like an onion: _____
   c. Blind spot; area in which there are no cones or rods: _____
   d. Area of sharpest vision; area of densest concentration of cones: _____
   e. Nonvascular, transparent, fibrous coat; most anterior eye structure: _____
   f. Layer containing neurons; if detached, causes blindness: _____
   g. Black layer; prevents reflection of light rays; also nourishes eyeball since it is vascular: _____
   h. A hole; appears black, like a circular doorway leading into a dark room: _____
   i. Regulates the amount of light entering the eye; colored part of the eye: _____
   j. Attaches to the lens by means of radially arranged fibers called the suspensory ligament: _____
   k. Jagged (serrated) margin of the retina: _____
   l. Located at the junction of iris and cornea; drains aqueous humor: _____

   Canal of Schlemm
   Choroid
   Ciliary muscle
   Cornea
   Fovea
   Iris
   Lens
   Optic disc
   Ora serrata
   Pupil
   Retina
   Sclera

7. In what ways are the formation, flow, and destination of aqueous fluid similar to that of cerebrospinal fluid?

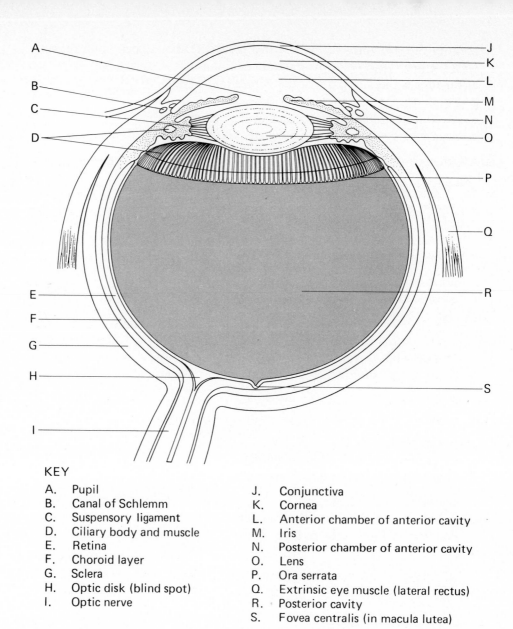

KEY

A. Pupil
B. Canal of Schlemm
C. Suspensory ligament
D. Ciliary body and muscle
E. Retina
F. Choroid layer
G. Sclera
H. Optic disk (blind spot)
I. Optic nerve

J. Conjunctiva
K. Cornea
L. Anterior chamber of anterior cavity
M. Iris
N. Posterior chamber of anterior cavity
O. Lens
P. Ora serrata
Q. Extrinsic eye muscle (lateral rectus)
R. Posterior cavity
S. Fovea centralis (in macula lutea)

Figure LG 18-3 Structure of the eyeball in horizontal section. Color as indicated in question 10 .

8. Where is vitreous humor located? What is its function?

9. Describe the conduction pathway for vision by arranging these structures in sequence. Write the letters in correct order on the lines provided. Also, indicate the four points where synapsing occurs by placing an asterisk (*) between the letters.

12

___ ___ ___ ___ ___

B. Bipolar cells
C. Cerebral cortex (visual areas)

G. Ganglion cells
OC. Optic chiasma
ON. Optic nerve
OT. Optic tract
P. Photoreceptor cells (rods, cones)
T. Thalamus

10. Answer these questions about the visual pathway to the brain. It may be helpful to refer to Figure 18-11 in the text. |13|
    a. Hold your left hand diagonally up and to the left so you can still see it. Your hand is in the (*temporal? nasal?*) visual field of your left eye, and

    in the _____ visual field of your right eye.
    b. Due to refraction, the image of your hand will be projected on to the (*left? right?*) (*upper? lower?*) portion of the retinas of your eyes.
    c. All nerve fibers from these areas of your retinas reach the (*left? right?*) side of your thalamus and cerebral cortex.
    d. Damage to the right optic tract, right side of thalamus, or right visual cortex would result in loss of sight of the (*left? right?*) visual fields of each eye.

G. Auditory sensations and equilibrium   (pages 489–497)

1. Indicate the location of each of these structures in the three principal regions of the ear: external ear (E), middle ear (M), and inner ear (I). Then write a sentence describing each structure. |14|
    a. Pinna

    b. External auditory meatus

    c. Ceruminous gland

    d. Tympanic membrane

    e. Opening into eustachian (auditory) tube

f. Auditory ossicles

g. Ligaments associated with ossicles

h. Oval and round windows

i. Bony and membranous labyrinth

j. Organ of Corti

k. Otoliths

l. Perilymph

2. Describe the auditory process by listing the sequence of events which occur from the time a friend calls out your name until you hear it, that is, until your brain perceives it. Include events in the external ear, middle ear, vestibule, cochlea, cochlear nerve and brain.

3. Check your understanding of the pathway of fluid conduction by placing the following structures in correct sequence. Write the letters on the lines provided.

15

—— —— —— —— —— —— —— ——

B. Basilar membrane
C. Cochlear duct (endolymph)
O. Oval window

RM. Reissner's membrane
RW. Round window
ST. Scala tympani (perilymph)
SV. Scala vestibuli (perilymph)

4. Contrast *static equilibrium* with *dynamic equilibrium*.

5. Explain the role of the *vestibule* in letting you know the position of your head, even if your eyes are closed.

6. Suppose you are riding on a winding, fast-moving roller coaster. Explain how your semicircular canals inform you of movements in forward, up-and-down, and circular directions.

H. Applications to health; key medical terms

16 1. Match the name of the disorder with the description.

a. Inflammation of sebaceous glands at the base of hair follicles of eyelashes:

_____

b. Excessive intraocular pressure resulting in blindness _____

c. Ringing in the ears: _____

_____

d. Crossed eyes: _____

e. Middle ear infection: _____

_____

f. Associated with vitamin A deficiency:

_____

g. Pinkeye: _____

h. A disturbance of the inner ear; cause unknown: _____

i. Loss of transparency of the lens:

_____

j. Farsightedness due to loss of elasticity of lens, especially after age 40:

_____

k. Condition requiring corrective lenses to focus distant objects: _____

_____

Cataract
Conjunctivitis
Glaucoma
Ménière's disease
Myopia
Night blindness
Otitis media
Presbyopia
Strabismus
Sty
Tinnitus

2. Discuss the following disorders.
   a. Labyrinthine disease

   b. Motion sickness

3. Describe how each of these disorders may be improved by drugs, surgery, or other treatment.
   a. Cataract

   b. Glaucoma

19
The Endocrine System

You have studied how coordination and integration of many body functions are accomplished by rapid-firing nerve impulses. In this chapter you will learn about a slower acting, yet equally vital, control system involving chemical messengers called hormones. You will first define the glands which produce hormones—endocrine glands (Objectives 1–2). You will learn about the major endocrine glands: their locations, the hormones they produce, and the role of each hormone in maintaining homeostasis (3–1.3). You will also study disorders and key medical terms related to the endocrine system (14–15).

Topics Summary

A. Endocrine glands
B. Pituitary (hypophysis)
C. Thyroid and parathyroids
D. Adrenals (suprarenals)
E. Other endocrine glands: pancreas, ovaries and testes, pineal (epiphysis cerebri), thymus
F. Applications to health; key medical terms

Objectives

1. Describe the relationship between the endocrine system and the nervous system in maintaining homeostasis.
2. Define an endocrine gland and an exocrine gland.
3. Describe the location, histology, and blood and nerve supply of the pituitary gland.
4. Explain how the pituitary gland is structurally and functionally divided into an adenohypophysis and neurohypophysis.
5. Discuss how the pituitary gland and hypothalamus are structurally and functionally related.
6. Describe the location, histology, and blood and nerve supply of the thyroid gland.
7. Describe the location, histology, and blood and nerve supply of the parathyroid glands.
8. Describe the location, histology, and blood and nerve supply of the adrenal glands.
9. Explain the subdivisions of the adrenal glands into cortical and medullary portions.
10. Describe the location, histology, and blood and nerve supply of the pancreas.
11. Explain why the ovaries and testes are classified as endocrine glands.
12. Describe the location, histology, and blood and nerve supply of the pineal gland.
13. Describe the location, histology, and blood and nerve supply of the thymus gland.
14. Discuss the symptoms of the following endocrine disorders: pituitary dwarfism, giantism, acromegaly, diabetes insipidus, cretinism, myxedema, exophthalmic goiter, simple goiter, tetany, osteitis fibrosa cystica, Addison's disease, Cushing's syndrome, aldosteronism, adrenogenital syndrome, gynecomastia, pheochromocytomas, diabetes mellitus, and hyperinsulinism.
15. Define key medical terms associated with the endocrine system.

274   **Learning Activities**

(page 504)   **A. Endocrine glands**

1. The two control systems of the body are the nervous system and the endocrine system.
   a. Compare the ways in which the nervous system and the endocrine system exert control over the body.

☐ 1   b. Stated very simply, what is the primary function of both systems?

2. Referring to Figure 19-1 in the text, identify the major endocrine glands. Approximate their locations in your own body.
   a. Compare the arrangement of the endocrine system with that of the nervous system.

☐ 2   b. How does the endocrine system achieve such widespread effects?

3. Complete the table contrasting *exocrine* and *endocrine* glands.

	Products Secreted into:	Examples
a. Exocrine		
b. Endocrine		

4. Describe the relationship between a hormone and its target cells or target organ.

5. Before you begin your study of different hormones, look at Table LG 19-1. As you proceed through the chapter, test your knowledge by filling in names, abbreviations, sources, and functions of these hormones. By the end of the chapter, you should be able to complete the table.

Table LG 19-1
Summary of Hormones and Regulating Factors

Abbreviation (if any)	Name	Source	Function
1. HGH or STH		Adenohypophysis	
2.	Thyroid-stimulating hormone		
3. ACTH			
4. FSH			
5. LH (ICSH)			
6.			Increases skin pigmentation
7.	Prolactin		
8. —	Oxytocin		
9. ADH			
10. —	Thyroxin		
11.			Decreases blood calcium
12.		Parathyroids	
13. —	Aldosterone		

Table LG 19-1 (Continued)

Abbreviation (if any)	Name	Source	Function
14. —	Cortisol, cortisone		
15. —	Sex hormones		
16.		Adrenal medulla	
17. —		Pancreas, alpha cells	
18. —	Insulin		
19. —		Pineal gland	
20. —	Thymosin		

(pages 504–506) **B. Pituitary (hypophysis)**

1. Complete this exercise about the pituitary gland.

   a. It is also known as the _____ .

   b. Why is it called the "master gland"? How many hormones does it produce?

   c. Where is it located? What protects it?

   d. Draw a diagram of the gland (actual size). Label the *anterior lobe, posterior lobe,* and *infundibulum.*

2. Add to your diagram above a drawing of the blood vessels supplying the anterior lobe.

3. Complete this exercise about adenohypophysis secretion.

   a. *Regulating factors* are released by the _____ and travel via the hypophyseal portal system of vessels to the adenohypophysis. Draw arrows to show their pathway in the diagram you drew above.

   b. Regulating factors pass from the hypothalamus into the _____

   _____ plexus of capillaries. These drain into the

   vessels known as _____ veins. Both

   sets of vessels are located in the stalk (or _____

   _____) of the pituitary gland.

   c. At the base of the infundibulum, the portal veins form the

   _____ plexus where hypothalamic factors can affect adenohypophysis cells, regulating release of their hormones.

   d. Two types of hormones are released from acidophil cells of the adenohypophysis. Initials of those hormones are _____ and _____. Basophil cells produce five types of hormones. Initials of those hormones are _____, _____, _____, _____, and _____.

   e. Blood vessels named _____ veins carry anterior pituitary hormones away from that gland for distribution to the body. Most of these hormones are classified as tropic hormones,

   meaning that their target organs are _____

   _____. Their action is to cause these target endocrine glands to secrete their own hormones. For example, the tropic hormone adrenocorticotropic hormone (ACTH) stimulates the _____ to release hormones such as _____.

4. Defend or dispute this statement: "The posterior pituitary does not truly synthesize any hormones."

5. Neurohypophysis secretion is not controlled by regulating factors such as those affecting the adenohypophysis. Describe the role of nerves in neuro-hypophysis secretion.

6. Contrast the two lobes of the hypophysis in this table.

	Adenohypophysis	Neurohypophysis
a. Structure (type of tissue)		
b. Blood supply		
c. Number of hormones released		

7. Describe hormonal control of lactation (milk production and release).

8. Imagine that you have just finished running a mile on a hot day. Answer these questions.
   a. Tell why you might be dehydrated.

   b. Explain how a reduced urine output might help to compensate for loss of body fluids in sweat.

c. How does release of ADH contribute to fluid balance in this situation?

C. Thyroid and parathyroids   (pages 507–509)

1. Draw a diagram of the thyroid gland. Label the *lobes* and the *isthmus*.

2. Describe the histology of the thyroid gland.

3. Name the two chemicals which together are called *thyroid hormone*. Explain how they are formed, stored, and transported through the bloodstream.

4. List the main actions of *thyroxin*.

4

5. Do this exercise about the relationship between thyroid and parathyroid glands.

   a. Another hormone produced by the thyroid gland is _____

   _____, which regulates distribution of _____ ion. High levels of calcitonin result in (*high? low?*) calcium levels in blood

   with _____ calcium levels in bone, since calcitonin keeps calcium from leaving bones and entering blood.

   b. The main action of parathyroid hormone is to (*increase? decrease?*) blood levels of calcium.

   c. The parathyroids are small, round masses embedded in the (*anterior? posterior?*) of the thyroid.

(pages 510–512)   **D. Adrenals (suprarenals)**

1. Each adrenal gland has two parts: the adrenal _____

   and the adrenal _____ .

2. Draw a diagram of the adrenal gland (actual size). Label *cortex* and *medulla*. Then label the three zones of the cortex, and write next to each the types of hormones synthesized there.

3. Discuss the blood and nerve supply to the adrenals.

4. Write two functions each for the following adrenocorticoid hormones.
   a. Aldosterone

5. Explain the relationship between the adrenal medulla and the sympathetic nervous system.

6. Name the two principal hormones secreted by the adrenal medulla. Circle the name of the hormone which accounts for 80 percent of adrenal medulla secretion.

7. Describe the effects of epinephrine and norepinephrine.

E. Other endocrine glands: pancreas, ovaries and testes, pineal (epiphysis cerebri), thymus (pages 512–516)

1. Describe the location, size, and structure of the pancreas.

2. The *islets of Langerhans* are located in the _____.

   Twenty-five percent of the islet cells are _____ cells, secreting glucagon; 75 percent are _____ cells, secreting

   _____.

3. The chief function of insulin is to (*raise? lower?*) the blood glucose level, while glucagon has the opposite effect in that it (*raises? lowers?*) the level of glucose in the blood. $\boxed{5}$

4. Name a pair of endocrine glands found only in females:_____

   _____. Name a pair found only in males: _____

   _____. Hormones produced by testes and ovaries will be discussed in Chapter 23.

5. Describe the location and size of the pineal gland.

6. List three hormones which may be secreted by the pineal gland, and describe functions of each.

7. Answer these questions about the thymus gland.
   a. State the location of the thymus.

   b. Describe the structure of this gland.

8. Complete this exercise about the thymus gland.
   a. The thymus gland is involved primarily in _____.
   b. This gland affects lymphatic tissue. It allows (*B? T?*) cells to develop the ability to destroy invading (*antigens? antibodies?*).

   c. A hormone called _____ from the thymus gland

   seems to facilitate change of B cells into _____ cells

   which can then produce _____ against antigens.
9. Name hormones produced by each of the following endocrine glands.
   a. Kidneys

   b. Placenta

10. One other system which is known to produce hormones is the _____ system. Name four hormones produced by the walls of the stomach and small intestine.

7

F. Applications to health

(pages 516–518)

8

1. Match the disorder with the hormonal imbalance.

_____ a. Deficiency of HGH in child; slow bone growth

_____ b. Excess of HGH in adult; enlargement of hands, feet, and jaw-bones

_____ c. Deficiency of ADH; production of enormous quantities of "taste-less" (nonsugary) urine

_____ d. Deficiency of insulin; hyperglycemia and glycosuria (sugary urine).

_____ e. Deficiency of thyroxin in child; short stature and mental retardation

_____ f. Deficiency of thyroxin in adult; edematous facial tissues, lethargy

_____ g. Excess of thyroxin; protruding eyes, "nervousness," weight loss

_____ h. Deficiency of thyroxin due to lack of iodine; most common reason for enlarged thyroid gland

_____ i. Deficiency of PTH; decreased calcium in blood and fluids around muscles resulting in abnormal muscle contraction

_____ j. Deficiency of adrenocorticoids; increased K^+ and decreased Na^+ resulting in low blood pressure and dehydration

Ac. Acromegaly
Ad. Addison's disease
C. Cretinism
DI. Diabetes insipidus
DM. Diabetes mellitus
E. Exophthalmic goiter
M. Myxedema
P. Pituitary dwarfism
S. Simple goiter
T. Tetany

2. Mrs. Marshall has diabetes mellitus. In the absence of sufficient insulin, her cells are deprived of glucose. Explain why she experiences the following symptoms.
   a. Increased urine production, thirst

b. Weight loss

c. Ketosis

d. Atherosclerosis

3. Contrast *maturity-onset* diabetes with *juvenile-onset* diabetes with regard to age of onset, percentage of cases, and dependence on insulin.

Answers to Numbered Questions in Learning Activities

1. Maintain homeostasis.

2. Hormones travel through blood which does reach virtually every part of the body.

3. (a) Hypothalamus. (b) Primary, hypophyseal portal, infundibulum. (c) Secondary. (d) HGH, PR; TSH, ACTH, FSH, LH, MSH. (e) Anterior hypophyseal, other endocrine glands, adrenal cortex, cortisol.

4. (a) Thyrocalcitonin (TCT or calcitonin), Ca^{2+}, low, high. (b) Increase. (c) Posterior.

5. Lower, raises.

6. (a) Immunity. (b) T, antigens. (c) Thymosin, plasma, antibodies.

7. Digestive.

8. (a) P. (b) Ac. (c) DI. (d) DM. (e) C. (f) M. (g) E. (h) S. (i) T. (j) Ad.

20

The Respiratory System

In this chapter you will learn about the system that provides for gaseous exchange between the external environment and body cells. You first will study the passageways that lead to the lungs (Objectives 1–5) and then will consider the composition of the lungs themselves (7–9). You will compare the volumes and capacities of air exchanged during respiration (10). Then you will examine the control of respiration (11) and modified forms of respiration (12). Finally you will study clinical aspects of the respiratory system, including disorders, treatments, and key medical terms (13–17).

Topics Summary

A. Nose, pharynx, larynx
B. Trachea, bronchi, lungs
C. Air volumes exchanged
D. Control of respiration, modified respiratory movements
E. CPR, Heimlich maneuver, smoking
F. Applications to health; key medical terms

Objectives

1. Identify the organs of the respiratory system.
2. Compare the structure of the external and internal nose.
3. Differentiate the three anatomical subdivisions of the pharynx and describe their roles in respiration.
4. Identify the anatomical features of the larynx related to respiration and voice production.
5. Describe the location and structure of the tubes that form the bronchial tree.
6. Contrast tracheostomy and intubation as alternative methods for clearing air passageways.
7. Identify the coverings of the lungs and the gross anatomical features of the lungs.
8. Describe the structure of a bronchopulmonary segment and a lobule of the lung.
9. Explain the structure of the alveolar-capillary membrane and its function in the diffusion of respiratory gases.
10. Compare the volumes and capacities of air exchanged during respiration.
11. Explain how the medullary rhythmicity area, apneustic area, and pneumotaxic area control respiration.
12. Define coughing, sneezing, sighing, yawning, sobbing, crying, laughing, and hiccuping as modified respiratory movements.
13. List the basic steps involved in cardiopulmonary resuscitation (CPR).
14. Explain how the Heimlich maneuver is performed.
15. Describe the effects of pollutants on the epithelium of the respiratory system.
16. Define nasal polyps, hay fever, bronchial asthma, emphysema, pneumonia, tuberculosis, hyaline membrane disease, sudden infant death syndrome, and carbon monoxide poisoning as disorders of the respiratory system.
17. Define key medical terms associated with the respiratory system.

Learning Activities

A. Nose, pharynx, larynx

1. Explain how the respiratory and cardiovascular systems work together to accomplish gaseous exchange among the atmosphere, blood, and cells.

2. Define the three main processes of respiration.
   a. Ventilation

   b. External respiration

   c. Internal respiration

3. Name the structures of the nose that are designed to carry out each of the following functions.
   a. Warm, moisten, and filter air

   b. Sense smell

   c. Assist in speech

4. Contrast the *conchae* with the *meati* in the nose.

5. Define *epistaxis* and briefly discuss treatment for this condition.

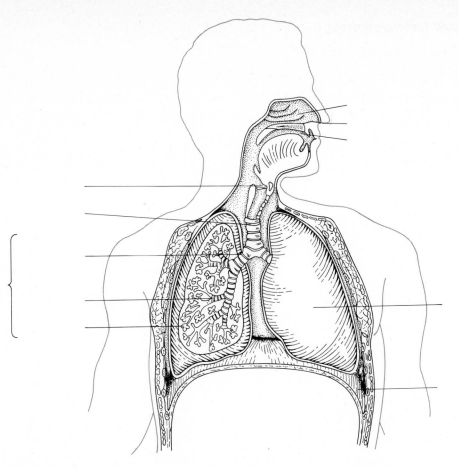

Figure LG 20-1 Organs of the respiratory system. Complete as directed.

6. On Figure LG 20-1 label the following structures: *external nares, internal nares, conchae, hard palate,* and *soft palate.*
7. State the functions of the pharynx.

8. On Figure LG 20-1, label the three portions of the pharynx. Write here which of these structures are located in each pharyngeal region: *adenoids, palatine tonsils, lingual tonsils, openings of fauces,* and *openings into eustachian (auditory) tubes.*
   a. Nasopharynx

   b. Oropharynx

c. Laryngopharynx

9. Describe the blood and nerve supply of the pharynx.

10. Describe each of these parts of the larynx.
   a. Epiglottis

   b. Glottis

   c. Thyroid cartilage

   d. Cricoid cartilage

   e. Arytenoid cartilages

11. Explain how the larynx prevents food from entering the trachea.

12. Tell how the larynx produces sound. Explain how pitch is controlled and what causes male pitch usually to be lower than female pitch.

13. After a larynx is removed (for example, due to cancer), what other structures help the laryngectomee to speak?

14. Define *laryngitis* and state two causes of it.

B. Trachea, bronchi, lungs (pages 529–539)

1. What is the function of each of these parts of the trachea?
   a. Ciliated pseudo-stratified columnar cells

   b. Goblet cells

   c. C-shaped cartilage rings

2. Define these terms and explain their clinical usage.
   a. Carina

   b. Tracheostomy

   c. Intubation

3. Describe the blood and nerve supply to the trachea.

4. The lower respiratory passageways are called the *tracheobronchial tree.*
   a. In what way do the passageways resemble a tree?

   b. On Figure LG 20-1 label the following: *trachea, primary bronchi, secondary bronchi, tertiary bronchi,* and *bronchioles.*

5. Describe the histological changes in cartilage, smooth muscle, and epithelium from primary bronchi to terminal bronchioles.

6. Explain how the absence of cartilage rings in bronchioles is significant during an asthma attack.

7. On Figure LG 20-1 color the *visceral pleura* red and the *parietal pleura* blue. Label the *pleural cavity.*

8. What is the function of serous fluid secreted by the pleura?

1   9. Write the correct term for each of these conditions.

   a. Inflammation of the pleura: _____

   b. Air in the pleural cavity: _____

   c. Blood in the pleural cavity: _____

2   10. Answer these questions about the lungs. (As you do the exercise, locate the parts of the lung on Figure 20-7 in your text.)

   a. The broad, inferior portion of the lung which sits on the diaphragm is

   called the _____. The upper narrow apex of each

   lung extends just superior to the _____. The costal

   surfaces lie against the _____.

   b. Along the mediastinal surface is the _____

   where the root of the lung is located. This root consists of

   _____.

c. Answer these questions with *right* or *left*. Which lung is thicker and broader? _____ In which lung is the cardiac notch located? _____ Which lung has just two lobes and so only two lobar bronchi? _____ Which lobe has a horizontal fissure? _____

11. There are an estimated 130,000 lobules in the lungs. Describe their structure by completing this exercise.

a. Air enters the lobule through a _____ bronchiole which then subdivides into a _____ bronchiole and into _____. These have tiny, balloon-shaped _____ in their walls (see Figure 20-9a in your textbook). Each is about 0.3 mm in diameter.

b. It is estimated that there are more than 2,000 alveoli in each lobule, or about _____ in the lungs. They provide a total surface area about the size of a tennis court, roughly _____ m^2. (A single sphere the size of the lung would have a surface area of only about 1/100 m^2!)

c. Each lobule contains a vessel named a(n) _____ which brings in pulmonary blood (*high? low?*) in oxygen. The arteriole leads to a network of _____ which surround the alveoli. After blood is oxygenated, it enters a pulmonary _____; this carries blood out of the lobule. Another type of vessel, a _____, also supplies each lobule.

d. Alveolar walls consist of two layers. The layer lining the alveolus consists of two main types of cells. Type I cells, which are _____ epithelium, compose most of the membrane. Type II cells produce an important substance called _____ which helps prevent collapse of alveoli. The function of a third type of cell is _____ _____.

e. This alveolar epithelial layer is very (*thick? thin?*). It sits on a thin layer known as a _____ membrane.

f. Blood capillaries consist of two layers, an _____ layer situated on a _____ membrane.

g. Diffusion of gases between alveoli and blood occurs across the _____ membrane. The four layers of this membrane, beginning from inside the alveolus, are _____, _____, _____, and _____ _____.

C. Air volumes exchanged

1. Each minute the average adult takes _____ breaths (respirations). Check your own respiratory rate and write it here: _____ breaths/minute.

2. Explain how a *spirometer* is used to measure volumes of air exchanged.

3. Match the lung volumes and capacities with the descriptions given. You may find it helpful to refer to Figure 20-10 in the text.

4

_____ a. The amount of air taken in with each inspiration during normal breathing is called _____ .

_____ b. At the end of a normal expiration, the volume of air left in the lungs is called _____ . Emphysemics who have lost elasticity of their lungs cannot exhale adequately, so this volume will be large.

_____ c. Forced exhalation can remove some of the air in FRC. The maximum volume of air which can be expired beyond normal expiration is called _____ . This volume will be small in emphysema patients.

_____ d. Even after the most strenuous expiratory effort, some air still remains in the lungs; this amount, which cannot be removed voluntarily, is called _____ .

_____ e. The volume of air which represents a person's maximum breathing ability is called _____ . This is the sum of ERV, TV, and IRV.

_____ f. Adding RV to VC gives _____ .

_____ g. The excess air a person can take in after a normal inhalation is called _____ .

_____ h. IRV + TV = _____ .

ERV. Expiratory reserve volume

FRC. Functional residual capacity

IC. Inspiratory capacity

IRV. Inspiratory reserve volume

RV. Residual volume

TLC. Total lung capacity

TV. Tidal volume

VC. Vital capacity

4. Indicate normal volumes for each of the following.

   a. TV = _____ ml (about _____ quart)

   b. TLC = _____ ml (_____ liters)

   c. VC = _____ ml

   d. ERV = _____ ml

   e. RV = _____ ml

D. Control of respiration, modified respiratory movements   (pages 541–543)

1. State the location and function of each of these respiratory control areas.
   a. Medullary rhythmicity area

   b. Apneustic area

   c. Pneumotaxic area

2. Define each of these terms.
   a. *Eupnea*

   b. *Costal breathing*

   c. *Diaphragmatic breathing*

3. A slight increase in the CO_2 level of the blood will tend to (*stimulate? inhibit?*) the inspiratory center.

6

E. CPR, Heimlich maneuver, smoking   (pages 543–548)

1. What do the letters CPR stand for? Describe briefly how this procedure is performed.

2. The procedure of *exhaled air ventilation* is sometimes called
_____. Explain how mouth-to-mouth resuscitation is
performed.

3. The air you inhale is about _____ percent oxygen. The air you exhale is
normally about _____ percent oxygen. This (*is? is not?*) sufficient oxygen to
help a person receiving this exhaled air via mouth-to-mouth resuscitation.

4. Name three common errors which should be avoided in giving mouth-to-
mouth resuscitation.

5. If two rescuers are performing CPR, one should apply at least _____ cardiac compressions per minute while the other exhales into the victim's
mouth once every _____ compression.

6. How does the *Heimlich maneuver* help to remove food which might other-
wise cause death by choking?

7. A host of effects upon the respiratory system are associated with smoking.
Describe these two.
   a. Bronchiogenic carcinoma

   b. Emphysema

1. Match the condition with the correct description.

_____ a. Permanent inflation of lungs due
to loss of elasticity; rupture and
merging of alveoli, followed by
their replacement by fibrous tissue

_____ b. Allergy reaction to plant and
other pollens

_____ c. Acute infection or inflammation
of alveoli which fill with fluid

_____ d. Spasms of small passageways
with wheezing and dyspnea

_____ e. The "most killing" communica-
ble disease

_____ f. Difficult, painful breathing

_____ g. Oxygen starvation

A. Asphyxia
B. Bronchial asthma
D. Dyspnea
E. Emphysema
H. Hay fever
P. Pneumonia
T. Tuberculosis

2. Discuss:
   a. Hyaline membrane disease

   b. Sudden infant death syndrome

3. Define these medical terms.
   a. Apnea

   b. Bronchitis

    c. Hypoxia

    d. Pneumothorax

    e. Pulmonary edema

    f. Pulmonary embolism

    g. Rales

Answers to Numbered Questions in Learning Activities

1. (a) Pleurisy. (b) Pneumothorax. (c) Hemothorax.

2. (a) Base, clavicles, ribs. (b) Hilus, bronchi, pulmonary vessels and nerves. (c) Right, left, left, right.

3. (a) Terminal, respiratory, alveolar ducts, alveoli. (b) Close to 300 million, 70. (c) Arteriole, low, capillaries, venule, lymphatic. (d) Simple squamous, surfactant, phagocytosis. (e) Thin, basement. (f) Endothelial, basement. (g) Respiratory, alveolar epithelium, basement membrane of alveolus, basement membrane of capillary, capillary endothelium.

4. (a) TV. (b) FRC. (c) ERV. (d) RV. (e) VC. (f) TLC. (g) IRV. (h) IC.

5. (a) 500, 0.5. (b) 6,000, 6. (c) 4,800. (d) 1,200. (e) 1,200.

6. Stimulate.

7. (a) E. (b) H. (c) P. (d) B. (e) T. (f) D. (g) A.

21

The Digestive System

Food is vital to homeostasis since food supplies the building blocks for all structures and the energy for various functions in the body. In this chapter you will study the system that changes complex foods into molecules which the body can utilize. You will define digestion and survey the general structure of digestive organs and their peritoneal coverings (Objectives 1–4). You will learn about the structure of each organ in the system: mouth, teeth, and salivary glands (5–7); stomach (8); pancreas, liver, and gallbladder (9–11); small intestine (12–13); and large intestine (14–15). You will consider some common disorders and key medical terms associated with the digestive system (16–21).

Topics Summary

A. General organization
B. Mouth, pharynx, esophagus
C. Stomach
D. Accessory organs: pancreas, liver, gallbladder
E. Small intestine
F. Large intestine
G. Applications to health; key medical terms

Objectives

1. Define digestion as a chemical and mechanical process.
2. Identify the organs of the gastrointestinal tract and the accessory organs of digestion.
3. Describe the structure of the wall of the gastrointestinal tract.
4. Define the mesentery, mesocolon, falciform ligament, lesser omentum, and greater omentum.
5. Describe the structure of the mouth and its role in mechanical digestion.
6. Identify the location and histology of the salivary glands and define the composition and function of saliva.
7. Identify the parts of a typical tooth and compare deciduous and permanent dentitions.
8. Describe the location, anatomy, and histology of the stomach and compare mechanical and chemical digestion.
9. Describe the location, structure, and histology of the pancreas.
10. Define the position, structure, and histology of the liver.
11. Describe the structure and histology of the gallbladder.
12. Describe those structural features of the small intestine that adapt it for digestion and absorption.
13. Describe the mechanical movements of the small intestine and define absorption.
14. Describe those structural features of the large intestine that adapt it for absorption and feces formation and elimination and describe the mechanical movements of the large intestine.
15. Describe the processes involved in feces formation and discuss the mechanisms involved in defecation.
16. List the causes and symptoms of dental caries, periodontal disease, and peptic ulcers.
17. Explain cirrhosis and gallstones as disorders of accessory organs of digestion.
18. Explain the causes and symptoms of appendicitis, diverticulitis, and peritonitis.
19. Compare the location of tumors of the gastrointestinal tract.
20. Describe the symptoms of anorexia nervosa.
21. Define key medical terms associated with the digestive system.

Learning Activities

(pages 554–558)   **A.** General organization

    1. Explain why food is vital to life. Give three specific examples of uses of foods in the body.

    2. List the five basic activities of the digestive system.

    3. Contrast *mechanical digestion* and *chemical digestion.*

    4. Identify the organs of the digestive system in Figure LG 21-1. Visualize the locations of these organs on yourself. Relate these to the nine abdominal regions (Figure LG 1-1).

    5. List the organs of digestion that are:
       a. Part of the gastrointestinal (GI) tract

       b. Accessory organs

    6. Refer to Figure LG 21-2, which shows a section of the wall of the GI tract. Label the four layers. Then briefly describe the structure and function of each layer.

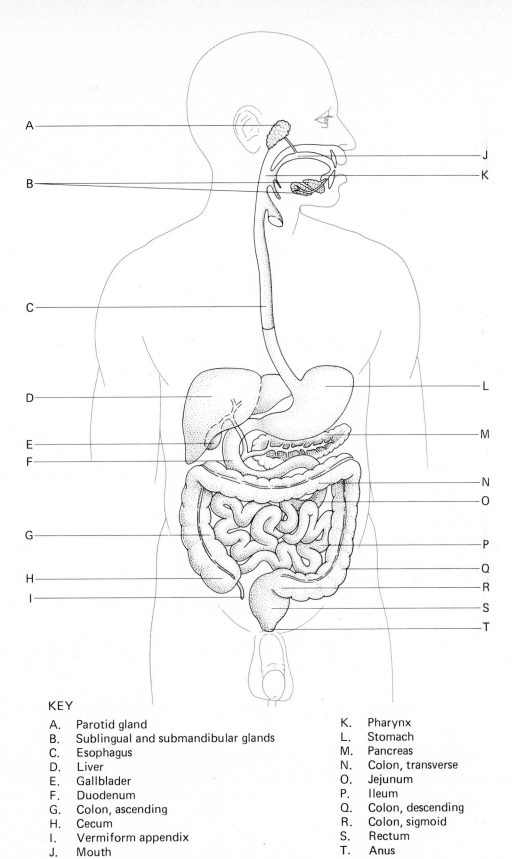

KEY

A. Parotid gland
B. Sublingual and submandibular glands
C. Esophagus
D. Liver
E. Gallblader
F. Duodenum
G. Colon, ascending
H. Cecum
I. Vermiform appendix
J. Mouth

K. Pharynx
L. Stomach
M. Pancreas
N. Colon, transverse
O. Jejunum
P. Ileum
Q. Colon, descending
R. Colon, sigmoid
S. Rectum
T. Anus

Figure LG 21-1 Organs of the digestive system.

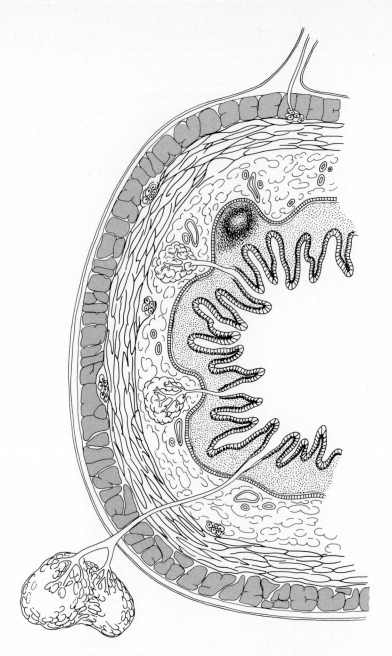

Figure LG 21-2 Gastrointestinal tract seen in cross section. Label as directed.

7. Contrast the epithelium of the tunica mucosa in the mouth and esophagus with that in stomach and intestine.

8. Is the mucosa of the GI tract keratinized? _____ State one advantage of this fact and one disadvantage of it.

9. Contrast *visceral peritoneum* and *parietal peritoneum*. Name the space between these two layers.

10. In what two ways does the peritoneum differ from the other serous membranes of the body, the pericardium and pleura?

1

11. Match the names of these peritoneal extensions with the correct descriptions.

2

_____ a. Attaches liver to anterior abdominal wall

_____ b. Binds intestines to posterior abdominal wall; provides route for blood and lymph vessels and nerves to reach small intestine

_____ c. Binds part of large intestine to posterior abdominal wall

_____ d. "Fatty apron"; covers and helps prevent infection in small intestine

_____ e. Suspends stomach and duodenum from liver

F. Falciform ligament
G. Greater omentum
L. Lesser omentum
M. Mesentery
Meso. Mesocolon

12. Note how extensive the peritoneal membrane is. Define *peritonitis* and discuss its clinical significance.

B. Mouth, pharynx, esophagus

1. Describe the following structures which form the oral cavity.
   a. Cheeks and lips

   b. Vestibule and oral cavity proper

   c. Palates, arches, and fauces

2. Describe these parts of the tongue.
   a. Extrinsic muscles

   b. Intrinsic muscles

   c. Frenulum

   d. Papillae

   e. Taste zones

3. Explain the problems that occur if the lingual frenulum is too short.

4. Describe the blood and nerve supply of the tongue.

5. Name the three salivary glands. Identify these glands on Figure LG 21-1 and visualize their locations on yourself.

6. Identify the location at which the duct of each salivary gland empties into the oral cavity.

7. How much saliva does the average adult produce each day? _____
_____

8. Name the chemical component of saliva that accomplishes each of the following functions.

   a. Destroys bacteria and so protects the mouth from infection: _____

   b. Composes over 99 percent of saliva; a medium for dissolving foods: _____

   c. Activates salivary amylase: _____

   d. Lubricates food so it can form a bolus: _____

3

9. Contrast the secretions of the three pairs of salivary glands according to the following criteria: thickness, whether secretions are serous or mucous, and presence of salivary amylase.

10. Describe the functions of *salivary amylase.*

11. Complete the diagram of a typical molar tooth in Figure LG 21-3. First, label *crown, cervix (neck)*, and *root* regions. Next, add to the diagram and label the following parts: *enamel, cementum, dentin, pulp cavity, root canal,* and *apical foramen.* Write a brief description of each structure next to its label. Finally, add blood vessels and nerves, the *periodontal ligament,* and surrounding bone.

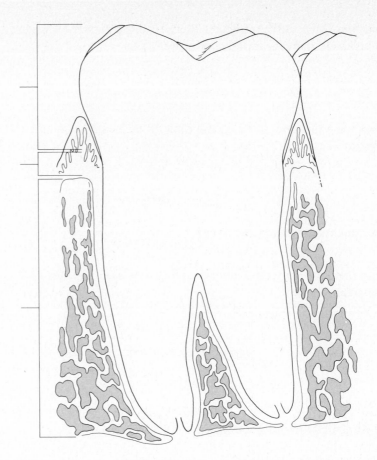

Figure LG 21-3 Outline of a typical molar tooth. Complete the diagram and label as directed.

12. State the clinical significance of *pyorrhea.*

13. Look at your own teeth in a mirror. Identify each of the following surfaces of your teeth: labial, buccal, lingual, palatine, mesial, distal, and occlusal.

14. Look at your own teeth in a mirror. Identify the different types of teeth and consider how the structural design of each relates to its function. Then answer these questions. (You may find it helpful to refer to Figure 21-8 in the text.)

4

   a. How many teeth are in a complete permanent dentition? _____ How many teeth do you have? _____ How many would be in a complete child's dentition, as in a 4-year-old? _____

   b. The four centrally located teeth are named _____. Lateral to these are _____. Posterior to the cuspid teeth are _____ and finally _____ teeth.

c. How many cuspids are in an adult set? _____ Premolars? _____

Molars? _____

d. The first permanent tooth to erupt is the _____ ; the

last is the _____ .

15. Another term for swallowing is _____ . Describe the process, explaining the role of each of these structures in deglutition: *tongue, oropharynx, soft palate, epiglottis,* and *esophagus.* (Refer to Figure 21-9 in your text.)

16. The esophagus connects the _____ to the

_____ .

17. Define each of these terms and relate it to the esophagus.
    a. *Peristalsis*

    b. *Sphincter*

C. Stomach (pages 569–575)

1. Describe the location of the stomach. Try to visualize its position on yourself. Determine its relationships to other organs.

2. On Figure LG 21-1, identify these regions of the stomach: *cardia, fundus, body,* and *pylorus.* Which is more lateral and inferior in location? (*Greater curvature? Lesser curvature?*) To which curvature is the greater omentum

attached? _____ The lesser omentum? _____

_____

3. Describe the *pyloric valve* according to location and function.

Contrast *pylorospasm* and *pyloric stenosis.*

4. Complete this table about gastric secretions.

Name of Cell	Type of Secretion	Function of Secretion
a. Chief (zymogenic)		
b.		Activates pepsinogen
c.	Mucus and intrinsic factor	

5. How does the muscularis of the stomach differ from this layer in the walls of other digestive organs?

6. Describe the blood and nerve supply to the stomach.

7. Explain how food is mixed in the stomach.

In which regions are mixing waves predominant? (*Fundus? Body?*)

8. Answer these questions about chemical digestion in the stomach.
   a. The most important enzyme released by the stomach is
      _____ . Gastric cells produce the enzyme in the in-
      active state, called _____ , which is activated by
      _____ .
   b. Pepsin is most active at very (*acid? alkaline?*) pH.
   c. State two factors which enable the stomach to digest protein without di-
      gesting its own cells (which are composed largely of protein).

   d. If mucus fails to protect the gastric lining, the condition known as
      _____ may result.
   e. Another enzyme produced by the stomach is _____
      which digests _____ . In adults it is quite (*effective?*
      ineffective?). Why?

9. Explain how chyme is propelled into the duodenum. Use these terms in
   your description: *peristalsis* and *pressure gradients.*

10. Food stays in the stomach for about _____ hours. Which food type leaves
    the stomach most quickly? _____ Which type stays in
    the stomach longest? _____
11. The stomach is responsible for (*much? little?*) absorption of foods. What
    types of substances are absorbed by the stomach?

D. **Accessory organs: pancreas, liver, gallbladder**   **(pages 575–584)**

1. Study Figures LG 21-1 and LG 21-4. Then complete these statements about
   the pancreas.

   a. The pancreas lies posterior to the _____ .

7

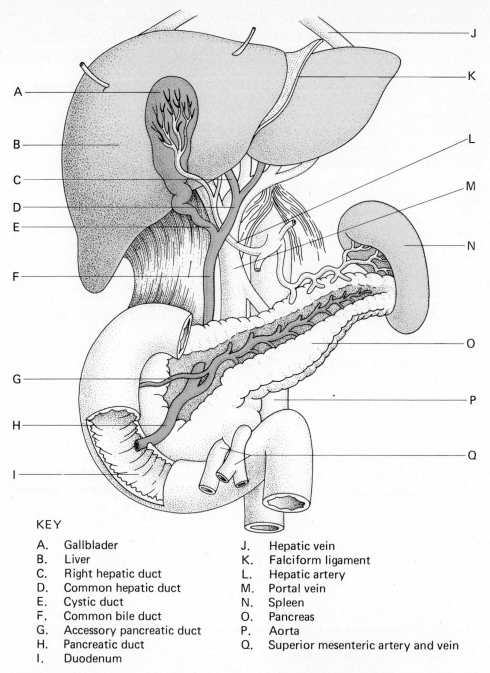

KEY

A. Gallbladder
B. Liver
C. Right hepatic duct
D. Common hepatic duct
E. Cystic duct
F. Common bile duct
G. Accessory pancreatic duct
H. Pancreatic duct
I. Duodenum

J. Hepatic vein
K. Falciform ligament
L. Hepatic artery
M. Portal vein
N. Spleen
O. Pancreas
P. Aorta
Q. Superior mesenteric artery and vein

Figure LG 21-4 Liver, gallbladder, pancreas, and duodenum, with associated blood vessels and ducts. (Stomach has been removed). Complete as indicated.

b. The pancreas is shaped roughly like a fish, with its *head* in the curve of the _____ and its *tail* nudging up next to the _____ .

c. The pancreas contains two types of glands. Its digestive juices are (*endocrine? exocrine?*) secretions; these pass into a duct which empties into the _____ .

d. Its endocrine portions are known as the _____ . These cells secrete two hormones, _____ and

_____. Typical of all hormones, these pass into (*ducts? blood vessels?*), specifically into veins which empty into the

_____ vein. The function of these hormones is to

control the level of _____.

2. Discuss the innervation of the following parts of the pancreas.
   a. Glands

   b. Blood vessels

3. Answer these questions about the liver. 　　　　　　　　　8

   a. This organ weighs about _____ kg (_____ lb). It lies in the

   _____ quadrant of the abdomen. Of its _____ lobes,

   the _____ is the largest.

   b. The _____ ligament separates the right and left lobes. In the edge of this ligament is the ligamentum teres, which is the

   obliterated _____ vein.

   c. Blood enters the liver via vessels named _____ and

   _____. In liver lobules, blood mixes in channels

   called _____ before leaving the liver via vessels

   named _____. (If you had difficulty answering these questions, review portal circulation in Chapter 12.)

4. While blood is in sinusoids of the liver, hepatic cells have ample opportunity to act on this blood, to modify it, to add new substances to it. List here six important functions of the liver. Use these key words: *plasma proteins, phagocytosis, detoxification, metabolism, vitamins* and *minerals*, and *bile.*

5. Describe the formation, flow, and storage of bile. Identify structures on Figures LG 21-4 which carry bile. Color these green. Then color arteries red, veins blue, and the pancreatic duct yellow.

6. Define *rugae*. Name two organs where rugae are located.

(pages 584–589) **E. Small intestine**

1. Describe these aspects of the small intestine.

   a. Its average diameter is _____ cm (_____ inch); its average length is about _____ m (_____ ft).

   b. Name the three main segments and indicate the length of each segment.

2. The small intestine is designed for effective secretion and absorption by modifications of its wall. Describe these by completing the following table.

Structure	Part(s) of Wall Involved	Function(s)
a. Crypts of Lieberkühn		
b.	Submucosa	Secrete alkaline mucus which neutralizes gastric acid
c. Goblet cells		
d.	Fingerlike projections of plasma membrane of mucosal epithelial cells	

Structure	Part(s) of Wall Involved	Function(s)
e. Villi		
f.	Deep folds in mucosa and submucosa	

3. Contrast the two types of nerve plexuses in the small intestine: Auerbach's and Meissner's.

4. Digestion in the small intestine occurs with the aid of secretions from three sources. Describe these in the table.

Source	Secretion	Amount (Daily)	Function(s)
a.	Bile		
b. Pancreas			
c.	Succus entericus		

5. Discuss the role of red blood cells in bile formation.

6. Define *jaundice*.

9

7. Write the main steps in the digestion of each of the three major food types. (You may wish to fill in the names of enzymes that catalyze each step; if so, write these above arrows.)

a. Carbohydrates

Polysaccharides ⟶ _____ ⟶

_____ ⟶ Monosaccharides

b. Proteins

Proteins ⟶ _____ ⟶

_____ ⟶ _____

c. Lipids

Neutral fats ⟶ Emulsified fats ⟶ _____

8. Complete this table describing the action of digestive enzymes.

Name	Source(s)	Function(s)
a. Amylase	1. Salivary glands 2.	
b.		Digests sucrose into glucose and fructose
c.		Starts protein digestion
d. Trypsin and chymo-trypsin		
e.		Activates trypsinogen
f. Erepsin		
g. Lipase		

9. Explain why proteases such as trypsin and pepsin are secreted in inactive form, that is, as trypsinogen and pepsinogen.

10. Contrast the three types of movements of the small intestine:
    a. *Rhythmic segmentation*

    b. *Pendular movements*

    c. *Propulsive peristalsis*

11. From the above learning activity, you can see the three main products of digestion. About 90 percent of all absorption of nutrients takes place into the walls of the (*stomach? small intestine? large intestine?*). Products of carbohydrate and protein digestion enter (*blood capillaries? lymph capillaries?*).

12. Describe how fatty acids and glycerol are absorbed. Use these terms: *micelles, chylomicrons, lacteals,* and *thoracic duct.*

F. Large intestine (pages 589–594)

1. Identify the regions of the large intestine in Figure LG 21-1. Draw arrows to indicate direction of movement of intestinal contents.

2. The total length of the large intestine is about ———— m (———— ft). More than 90 percent of its length consists of the part known as the

   ——————————————— .

3. Describe each of the following structures.
   a. Ileocecal valve

   b. Appendix

c. Anal columns

d. Taeniae coli

e. Haustra

4. Define the following conditions. Explain what might cause each.
   a. Appendicitis

   b. Hemorrhoids

5. Contrast the arterial supply to each of these parts of the large intestine:
   a. Cecum and colon

   b. Rectum and anal canal

6. Contrast the parts of the digestive system innervated by the following portions of the parasympathetic nervous system:
   a. Vagus

   b. Pelvic splanchnic nerves

7. Write a sentence describing each of these types of movements of the large
   intestine.
   a. Haustral churning

  b. Peristalsis

  c. Mass peristalsis

8. Name the chemical components of feces.

9. Explain the role of bacteria in the large intestine.

10. Most water absorption occurs in the _____. The large intestine absorbs (*some? no?*) water.

10

11. Describe the process of defecation. Include these terms: *rectal receptors, rectal muscles, sphincters, diaphragm,* and *abdominal muscles.*

G. Applications to health; key medical terms   (pages 594–599)

1. Explain how tooth decay occurs. Include roles of *bacteria, plaque,* and *acid.* List the most effective known measures for preventing dental caries.

2. Briefly describe these disorders, stating possible causes of each.
   a. Periodontal disease

   b. Peptic ulcers

   c. Cirrhosis

   d. Gallstones

   e. Appendicitis

   f. Diverticulitis

   g. Anorexia nervosa

3. Explain why a patient may be given large doses of antibiotics prior to abdominal surgery.

22

The Urinary System

Digestion and metabolism of nutrients results in production of some wastes, such as carbon dioxide, water, and urea. In this chapter you will examine the urinary system, which eliminates wastes and controls composition and volume of body fluids. You will study the urine-producing organs, the kidneys: their structure and blood and nerve supply (Objectives 1–3). You will trace the pathway of urine through other urinary organs leading to the exterior of the body (4–5, 7). Finally, you will consider some disorders, treatment, and key medical terms associated with the urinary system (6, 8–11).

Topics Summary

A. Kidneys
B. Ureters, urinary bladder, urethra
C. Applications to health; key medical terms

Objectives

1. Identify the external and internal gross anatomical features of the kidneys.
2. Define the structural features of a nephron.
3. Describe the blood and nerve supply to the kidneys.
4. Describe the location, structure, and physiology of the ureters.
5. Describe the location, structure, histology, and function of the urinary bladder.
6. Compare the causes of incontinence, retention, and suppression.
7. Describe the location, structure, and physiology of the urethra.
8. Discuss the principle of hemodialysis.
9. Contrast the abnormal constituents of urine and their clinical significance.
10. Discuss the causes of ptosis, kidney stones, gout, glomerulonephritis, pyelitis, cystitis, nephrosis, and polycystic disease.
11. Define key medical terms associated with the urinary system.

Learning Activities

A. Kidneys

1. List the functions of the urinary system.

2. List three other systems that aid in elimination of wastes.

1

3. Identify the organs that make up the urinary system on Figure LG 22-1 and answer the following questions about them.

   a. The kidneys are located at about (*waist? hip?*) level, between _____ and

      _____ vertebrae.

   b. The kidneys are in an extreme (*anterior? posterior?*) position in the abdomen. They are described as _____ since they are posterior to the peritoneum.

4. What are the dimensions of the kidneys?

5. Describe the layers of tissue around each kidney.

6. Identify the parts of the internal structure of the kidney on Figure LG 22-1A–G. Then check your understanding of these structures by coloring the parts of the *cortex* red, *medulla* green, and *pelvis* yellow.

7. The functional unit of the kidney is called a _____ . Part of a nephron is shown in Figure LG 22-2. Label: *Bowman's capsule* (*visceral* and *parietal* layers) and the five parts of the *renal tubule* (*proximal convoluted tubule, descending limb of Henle, ascending limb of Henle, distal convoluted tubule,* and *collecting duct*).

8. Contrast the locations of the *cortical nephron* and the *juxtaglomerular nephron.* Note which parts of the nephrons lie in the cortex and which parts are in the medulla.

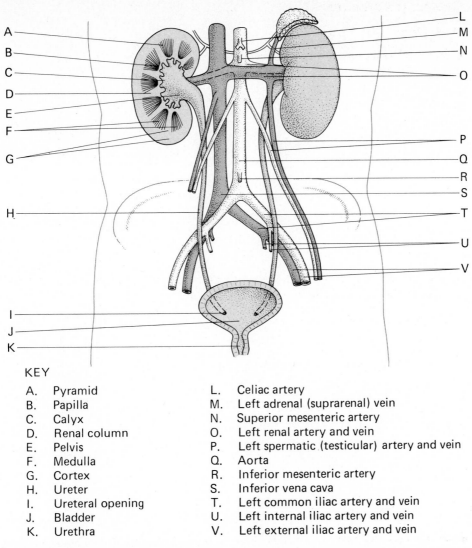

KEY

A.	Pyramid	L.	Celiac artery
B.	Papilla	M.	Left adrenal (suprarenal) vein
C.	Calyx	N.	Superior mesenteric artery
D.	Renal column	O.	Left renal artery and vein
E.	Pelvis	P.	Left spermatic (testicular) artery and vein
F.	Medulla	Q.	Aorta
G.	Cortex	R.	Inferior mesenteric artery
H.	Ureter	S.	Inferior vena cava
I.	Ureteral opening	T.	Left common iliac artery and vein
J.	Bladder	U.	Left internal iliac artery and vein
K.	Urethra	V.	Left external iliac artery and vein

Figure LG 22-1 *Left side of diagram:* Location of urinary system organs. Structures A to G are parts of the kidney. *Right side of diagram:* Blood vessels of the abdomen. Color as directed.

9. Name the structures which form a *renal corpuscle.*
       2

10. Describe the *endothelial-capsular membrane.*
   a. Name the layers composing it.

   b. List the functional advantages of these structures: *endothelial pores, pedicels,* and *filtration slits.*

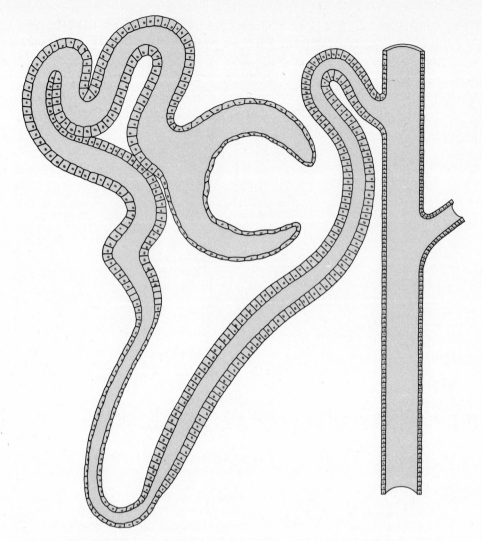

Figure LG 22-2 Diagram of a nephron (partial) to be completed as indicated.

11. List the five parts of a renal tubule and write next to each part the type of epithelium forming that portion of the tubule.

12. Describe the pathway blood takes as it courses through the kidneys.
    a. Name the vessels in order. (Consult Figure 22-6 in the text.)

b. On Figure LG 22-2, draw and label these vessels: *afferent arteriole, glomerular capillaries, efferent arteriole, peritubular capillaries, vasa recta,* and *veins.* (See Figure 22-3 in your text.)

13. Locate the *juxtaglomerular apparatus* on Figure 22-7 in the text.
    a. What structures form it?

    b. Draw and label it on Figure LG 22-2.

    c. Describe the function of the juxtaglomerular apparatus.

14. In what main ways is blood modified as it passes through the kidneys? ⬜3

B. Ureters, urinary bladder, urethra

(pages 614–619)

1. Refer to Figure LG 22-1 as you complete the following exercise. ⬜4

   a. Ureters connect _____ to _____.

   Ureters are about _____ cm (_____ inches) long.
   b. These tubes enter the urinary bladder at two of the three corners of the

   _____. The third corner marks the opening into the

   _____.

   c. The urinary bladder is located in the (*abdomen? pelvis?*). Two sphincters lie just inferior to it. The (*internal? external?*) sphincter is under voluntary control.
   d. Urine leaves the bladder via a tube named the _____

   _____. In females the length of this tube is about

   _____ cm; in males it is about _____ cm.

2. Ureters, urinary bladder, and urethra are all lined with _____

_____ membrane. What is the clinical significance of that fact?

3. Define *peristalsis* and describe the role of peristalsis in the function of ureters.

4. Describe the *micturition reflex*.

In this reflex (*sympathetic? parasympathetic?*) nerves stimulate the

_____ muscle of the urinary bladder and cause relaxation of the internal sphincter.

5. Define *incontinence*.

6. Contrast *retention* and *suppression*.

(pages 619–622) **C. Applications to health; key medical terms**

1. Explain how *hemodialysis* can help a person with impaired kidney function to maintain homeostasis.

   a. Define *dialysis*.

   b. Name the blood vessel that is usually connected to the hemodialysis device.

   5

   c. Tell what percentage of the patient's blood is in the machine at a time.

23

The Reproductive Systems

In all of the previous chapters you have studied homeostatic mechanisms that promote the health of the individual. Now you will consider how the human organism is adapted for continuity of the species. In this chapter you will define reproduction (Objective 1) and examine reproductive organs in the male (2–11) and in the female (12–20). You will learn about hormonally regulated cycles in the female (16). You will also discuss applications to health and key medical terms associated with reproduction (21–25).

Topics Summary

A. Male reproductive system

B. Female reproductive system: ovaries, uterine tubes, uterus

C. Female reproductive system: endocrine relations

D. Female reproductive system: vagina, external genitalia, mammary glands

E. Applications to health; key medical terms

Objectives

1. Define reproduction and classify the organs of reproduction by function.
2. Describe the structure and function of the scrotum.
3. Explain the structure, histology, and functions of the testes.
4. Describe the effects of testosterone.
5. Describe the straight tubules and rete testis as components of the duct system of the testes.
6. Describe the location, structure, histology, and functions of the ductus epididymis.
7. Explain the structure and functions of the ductus deferens.
8. Describe the structure and function of the ejaculatory duct.
9. List and describe the three anatomical subdivisions of the male urethra.
10. Explain the location and functions of the seminal vesicles, prostate gland, and bulbourethral glands, the accessory reproductive glands.
11. Explain the structure and functions of the penis.
12. Describe the location, histology, and functions of the ovaries.
13. Explain the location, structure, histology, and functions of the uterine tubes.
14. Describe the location and ligamentous attachments of the uterus.
15. Explain the histology and blood supply of the uterus.
16. Describe the effects of estrogens and progesterone.
17. Describe the location, structure, and functions of the vagina.
18. List and explain the components of the vulva and note the function of each component.
19. Describe the anatomical landmarks of the perineum.
20. Explain the structure and histology of the mammary glands.
21. Explain the symptoms and causes of disorders of the reproductive systems, including venereal diseases, trichomoniasis, nongonococcal urethritis, prostate disorders, impotence, infertility, and menstrual abnormalities.
22. Contrast ovarian cysts, endometriosis, and leukorrhea.
23. Discuss breast and cervical cancer. Explain special detection procedures such as mammography, thermography, ultrasonic scanning, and CT/M.
24. Describe pelvic inflammatory disease.
25. Define key medical terms associated with the reproductive systems.

   A. Male reproductive system

    1. Define:
       a. Gonad

       b. Gamete

   1   2. List the four main types of organs comprising the male reproductive system.

    3. Answer these questions about the scrotum.
       a. Describe the structure and contents of the scrotum.

       b. How does the temperature in the scrotum compare with that in the abdomen?

       c. Where do testes develop in early fetal life?

       d. Define *cryptorchidism*.

    4. Discuss these aspects of testes structure: size and weight, *lobules, seminiferous tubules, Sertoli cells*, and *interstitial cells of Leydig*.

5. Contrast the blood supply of the testes with that of the scrotum.

6. In the space below, draw a diagram of a sperm. Label: *head, middle piece, tail, nuclear material, acrosome,* and *mitochondria.*

7. Match each of the hormones listed below with descriptions.   2

   a. Released by the hypothalamus; high levels stimulate release of two anterior pituitary hormones (two answers): ___

      _____

      FSH
      FSHRF
      ICSH
      ICSHRF

   b. Stimulates seminiferous tubules to start sperm production:_____

   c. Stimulates release of testosterone: ___

      _____

   d. Produced by interstitial cells of Leydig:

      _____

8. List five functions of testosterone.

9. Carefully study structures in Figures 23-1 and 23-2 in the text. Then write in correct order (on the lines provided) the letters of structures in the pathway of sperm from their site of formation to point where they leave the body.   3

   A. Ductus (vas) deferens   F. Epididymis, tail
   B. Efferent ducts   G. Rete testis
   C. Ejaculatory duct   H. Seminiferous tubules
   D. Epididymis, body   I. Straight tubules
   E. Epididymis, head   J. Urethra

   ___ ___ ___ ___ ___ ___ ___ ___ ___ ___

10. Describe the structure, location, and functions of the epididymis.

4   11. Answer these questions about the ductus (vas) deferens.

     a. The ductus deferens is about _____ cm (_____ inches) long.

     b. It is located (*entirely? partially?*) in the abdomen. It enters the abdomen

        via the _____. Protrusion of abdominal contents

        through this weakened area is known as _____.

     c. What structures besides the vas compose the *seminal cord?*

     d. A vasectomy is performed at a point along the section of the vas that is

        within the _____. Must the peritoneal cavity be
        entered during this procedure? (*Yes? No?*) Will the procedure affect tes-
        tosterone level of the male? (*Yes? No?*) Why?

     e. Each ductus deferens empties its contents into the _____

        _____ duct which leads to the _____

        _____.

12. Name the parts of the male urethra. Identify these on Figure 23-1 in your
    text. Which portion is longest?

13. Describe the following structures according to location, structure, and function. (Refer to Figure 23-7 in your text.)

    a. Seminal vesicles

    b. Prostate gland

    c. Bulbourethral glands

14. Explain the clinical significance of the shape and location of the prostate gland.

    5

15. Answer the following questions about semen.

    a. The average amount per ejaculation is _____ ml.

    b. The average range of number of sperm per milliliter is _____

    _____ .

    c. List characteristics of semen which enhance fertility.

16. Refer to Figure 23-8 in the text and do this exercise.

    6

    a. The _____, or foreskin, is a covering over the

    _____ of the penis. Surgical removal of the foreskin

    is known as _____ .

    b. The names *corpus* _____ and *corpus* _____

    _____ indicate that these bodies of tissue in the penis contain spaces that distend in the presence of excess

    _____ . (*Sympathetic? Parasympathetic?*) stimulation causes dilation of blood vessels of the penis. As a result, blood is prohibited from leaving the penis, since (*arteries? veins?*) are compressed. This temporary state is known as (*erect? flaccid?*) state.

    c. The urethra passes through the corpus _____ . The

    urethra functions in the transport of _____ and

    _____ . During ejaculation, what prevents sperm from entering the urinary bladder and urine from entering the urethra?

17. Name the muscles of the penis and identify them on Figure 10–13 in your text.

(pages 637–646) **B.** Female reproductive system: ovaries, uterine tubes, uterus

1. Name the organs in the female reproductive system.

2. Describe the ovaries according to:
   a. Shape and size

   b. Position and ligaments

   c. Structure (layers and other parts)

   d. Functions

7   3. Complete this exercise about the uterine tubes.

    a. Uterine tubes are also known as _____ tubes. They are about _____ cm (_____ inches) long.

    b. The funnel-shaped open end of each tube is known as the _____ . Its fingerlike projections, called _____ , are close to (but not in direct contact with) the _____ .

    c. What structural features of the tube enhance passage of ova into and through the tubes?

d. List two functions of uterine tubes.

e. What is an ectopic pregnancy?

4. In the space below, draw the outline of a uterus (actual size and shape before first pregnancy). Label: *fundus, body,* and *cervix.*

5. Answer these questions about uterine position. 8

   a. The organ that lies anterior and inferior to the uterus is the

      _____ . The _____ lies posterior
      to it.

   b. The fundus of the uterus is normally tipped (*anteriorly? posteriorly?*) If it is malpositioned posteriorly, it would be (*anteflexed? retroflexed?*).

   c. The structures that hold the uterus in position are referred to as

      _____ . The _____ ligaments attach the uterus to the sacrum. The _____ ligaments pass through the inguinal canal and anchor into external genitalia (labia majora). The _____ ligaments are broad thin sheets of peritoneum extending laterally from the uterus.

      The _____ ligaments attach to the base of the uterus and are most important in preventing drooping (prolapse) of the uterus.

6. Where is the *rectouterine pouch,* or pouch of Douglas? (Refer to Figure 23-9 in your text.)

7. On your diagram of the uterus color these three layers: *endometrium,* red; *myometrium,* blue; *peritoneum* (*serosa*), yellow.

8. Name the two layers of the endometrium. Which is supplied by spiral arteries and is shed in menstruation?

9. Contrast the blood supply of each of the following structures:
   a. Ovaries

   b. Uterine tubes

   c. Uterus

(pages 646–648) **C. Female reproductive system: endocrine relations**

1. Contrast *menstrual cycle* and *ovarian cycle.*

2. List the functions of each of these female hormones.
   a. FSHRF

   b. LHRF

   c. FSH

   d. LH

   e. Estrogen

   f. Progesterone

3. Contrast *menarche, menopause,* and *climacteric.*

D. **Female reproductive system: vagina, external genitalia, mammary glands**   (pages 648–650)

1. Refer to Figure 23-9 in your textbook. In the normal position, at what angle does the uterus join the vagina? _____   9

2. Describe these vaginal structures.
   a. Fornix

   b. Rugae

   c. Hymen

3. The pH of vaginal mucosa is (*acid? alkaline?*). What is the clinical significance of this fact?

4. On Figure LG 23-1 label the following parts of the vulva: *mons pubis, labia majora, labia minora, clitoris,* and *vaginal orifice.*
5. Identify male homologues for each of the following.   10

   a. Labia majora: _____

   b. Clitoris: _____

   c. Lesser vestibular (Skene's) glands: _____

   d. Greater vestibular (Bartholin's) glands: _____

6. Name the structures which form the four corners of the *perineum* in both sexes. (Refer to Figure 23-18 in the text.)   11

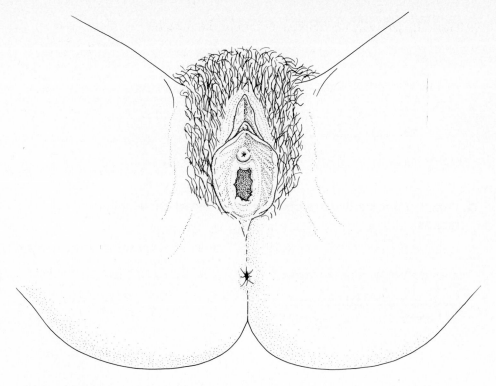

Figure LG 23-1 Vulva. Complete as directed.

Draw the borders of this diamond-shaped area on Figure LG 23-1. Then draw a line between the ischial tuberosities. The two triangles formed are

called the _____ and the _____ .

All of the female genitalia are located in the _____ triangle.

7. Locate the *clinical perineum* in your diagram. Draw a red line where the episiotomy incision would be performed.

8. Each mammary gland is divided into 15 or 20 compartments called

_____ . Describe the components of a lobule.

How does the size of a breast correlate to the amount of milk produced by it?

9. Trace the pathway of the duct system of the breasts.

10. Describe hormonal influence on breast development.

E. Applications to health; key medical terms　　　　　　　**(pages 650–656)**

1. Define *venereal disease.*

2. Describe the main characteristics of these venereal diseases by completing the table.

Disease	Causative Agent	Main Symptoms
a. Gonorrhea		
b.	*Treponema pallidum*	
c. Genital herpes		

Disease	Causative Agent	Main Symptoms
d.	*Trichomonas vaginalis*	
e. NGU		

3. Check your understanding of the following disorders by matching them with the correct description.

12

_____ a. Painful menstruation, partly due to contractions of uterine muscle

_____ b. Possible cause of blindness in newborns if bacteria transmitted to eyes during birth

_____ c. Spreading of uterine lining into abdominopelvic cavity via uterine tubes

_____ d. Caused by bacterium; may involve many systems in tertiary stage

_____ e. Absence of menstrual periods

_____ f. Benign tumor with firm, rubbery consistency; easily moved within breast

_____ g. Caused by flagellated protozoan

A. Amenorrhea
D. Dysmenorrhea
E. Endometriosis
F. Fibroadenoma
G. Gonorrhea
S. Syphilis
T. Trichomoniasis

4. Define these conditions and state possible causes of each.
   a. Impotence

   b. Infertility

5. Discuss roles of prostaglandins and progesterone in menstrual discomfort.

24

Developmental Anatomy

In the last chapter you learned about the systems involved in production of sperm and ova. In this chapter you will learn about the developmental processes which lead to the formation of a new individual. You will examine the special process of division which results in gametes (Objectives 1–2). You will consider the union of gametes and the implantation of the fertilized egg (3–4). You will study development through embryonic and fetal periods (5–8). You will discuss labor and birth, adjustments required of the newborn (9–10), the process of lactation (12), and possible effects on development of harmful environmental factors (11). Then you will study birth control (13), amniocentesis (14), and key medical terms associated with developmental anatomy (15).

Topics Summary

A. Gamete formation

B. Fertilization and implantation

C. Embryonic development

D. Developmental anatomy of body systems

E. Hormones of pregnancy, parturition and labor, adjustments at birth, lactation

F. Amniocentesis, birth control, key medical terms

Objectives

1. Define meiosis.
2. Contrast the events of spermatogenesis and oogenesis.
3. Compare the role of the male and female in sexual intercourse.
4. Explain the activities associated with fertilization and implantation.
5. Discuss the formation of the three primary germ layers, embryonic membranes, and placenta and umbilical cord as the principal events of the embryonic period.
6. Outline the principal developmental events associated with the major systems of the body.
7. List several body structures produced by the three primary germ layers.
8. Compare the sources and functions of the hormones secreted during pregnancy.
9. Explain the events associated with the three stages of labor.
10. Explain the respiratory and cardiovascular adjustments that occur in an infant at birth.
11. Discuss potential hazards to the embryo and fetus associated with chemicals and drugs, irradiation, alcohol, and cigarette smoking.
12. Discuss the mechanism of lactation.
13. Contrast the various kinds of birth control and their effectiveness.
14. Explain the procedure of amniocentesis and its value in diagnosing diseases in the newborn.
15. Define key medical terms associated with developmental anatomy.

A. Gamete formation

1. Define *developmental anatomy.*

2. Discuss how the *chromosome number* of a human epithelial cell differs from that of a human gamete. Contrast *diploid* (2n) with *haploid* (n).

3. By what process is the diploid number of a potential sperm cell converted to the haploid number? What is the comparable process in the ovary?

☐1   4. Complete this exercise about sperm formation.

    a. The process of spermatogenesis requires about 2½ (*hours? days? months? years?*). It occurs in _____ of testes under influence of the hormone _____. As cells proceed through the stages of this process, they move (*toward? away from?*) the lumen of the tubule.

    b. Cells that are precursors of all sperm are named _____ _____. They are products of previous mitotic divisions, and so they contain _____ chromosomes, the (*haploid? diploid?*) number for humans.

    c. As at the start of mitosis, meiosis begins with one replication of DNA. Cells then contain 46 doubled chromosomes (or 46 × 2 _____ _____) and are known as _____ _____.

    d. These chromosomes line up in _____ pairs. This means that, for example, in your own cells undergoing meiosis (regardless of your sex), a doubled paternal chromosome (one you originally received from your _____) lies next to a doubled maternal chromosome (one you received from your _____). Since all of your cells derived from that one original sperm and egg, all cells, including those in gonads, contain _____ homologous pairs of chromosomes. The four homologous chromatids in a homologous pair are together known as a _____. While in homologous pairs, members of the tetrad can exchange DNA with one another. This process is called _____.

e. The chromosomes now pull apart. As they do so, the tetrads are split in half. Depending on how members of homologous pairs had aligned themselves, chromosomes will be "assigned" to daughter cells of this first meiotic division. In other words, the two resulting cells will not be identical, as are daughter cells resulting from _____

_____ division. But as a result of homologous pairing, as well as _____, great variety in cells produced from this division will occur.

f. Cells resulting from the first meiotic division are called _____. These cells undergo a second meiotic division, producing cells called _____. These cells are then modified to form mature gametes called _____. Once formed, mature sperm (*do? do not?*) divide again.

5. Answer these questions about aspects of oogenesis which differ from spermatogenesis.
a. When do oogonia terminate mitosis?

b. How are polar bodies produced?

6. Contrast mitosis with meiosis by completing this table. $\boxed{2}$

Process	Identical or Different Resulting Cells	Number of Chromosomes in Resulting Cells	Number of Cells Resulting From One Cell
a. Mitosis			
b. Meiosis in spermatogenesis			
c. Meiosis in oogenesis			

B. Fertilization and implantation

(pages 662–668)

1. Define *sexual intercourse* or *coitus.*

2. Answer these questions about the process. $\boxed{3}$

a. (*Sympathetic? Parasympathetic?*) impulses from the (*brain stem? sacral cord?*) cause dilation of arteries of the penis, leading to the (*flaccid? erect?*) state.

b. (*Sympathetic? Parasympathetic?*) impulses cause sperm to be propelled into the urethra. This process is called (*emission? ejaculation?*). Expulsion from urethra to the exterior of the body is called _____ _____ .

3. Describe the processes of *erection, lubrication,* and *orgasm* in the female.

4. Most of the lubricating fluid which facilitates sexual intercourse is produced by the (*male? female?*). Which structures produce this fluid?

5. Answer these questions about the process of fertilization.
   a. How many sperm are ordinarily introduced into the vagina during sexual intercourse? About how many reach the area where the ovum is located?

   b. What two mechanisms may be involved in sperm transport within the female reproductive organs?

   c. Name the usual site of fertilization.

   d. What reproductive structure secretes hyaluronidase? What is the function of this enzyme?

   e. How many sperm fertilize an ovum? What prevents further sperm from entering the ovum?

6. Describe the roles of these structures in fertilization. Tell how many chromosomes are in each.
   a. Male pronucleus

   b. Female pronucleus

   c. Segmentation nucleus

   d. Zygote

7. Summarize the main events during the week following fertilization. Use these terms: *cleavage, blastomere, morula,* and *blastocyst.* (See Figures 24-5 and 24-8 in your text.)

8. Draw a diagram of a blastocyst. Label: *blastocoel, inner cell mass,* and *ectodermal cells* (*trophectoderm*). Color red the cells which will become the embryo. Color blue the cells which will become part of the placenta.

9. The blastocyst becomes attached to the endoderm _____ days after fertilization, which is about day _____ of the monthly cycle. This process is called _____. Notice that the developing embryo is implanted in the lining of the uterus even before the first day of the next menstrual cycle is "missed."

10. Discuss the significance of external human fertilization.

(pages 668–775) **C. Embryonic development**

1. Contrast *embryo* with *fetus.*

2. Complete this exercise about the embryonic period. (Refer to Figure 24-7 in the text.)

    a. Between the trophectoderm and the inner cell mass, a cavity forms; this is the _____ cavity.

    b. The inner cell mass forms the embryonic _____. It consists initially of two primary germ layers. The upper cells (closer to the site of implantation), called _____-derm cells, surround the _____ cavity. These cells eventually fold downward so that the amniotic cavity completely surrounds the fetus. Ectodermal cells then form both the outer tissue of the fetus, that is, the _____, and also the fetal membrane named the _____. This delicate structure is the "_____" surrounding the fetus which is broken at birth.

    c. The lower cells of the embryonic disc, called _____ cells, line the primitive _____, and for a while extend out to form the _____ sac. This sac (*is? is not?*) as vital in humans as it is in birds.

d. Between the endoderm and the ectoderm, a third germ layer forms dur-

ing the third week of development. This is the _____

_____ . Some of these cells form fetal structures, and

some of them move around to line the original trophectoderm cells (Fig-

ure 24-7c). Together these cells form part of the fetal membrane called

the _____ . The cells surround the original _____

_____ cavity, now called the _____

_____ . This space will exist between the amnion

and the chorion (Figure 24-10 in the text), with the _____

_____ the more superficial of the two membranes.

3. A fourth fetal membrane (in addition to chorion, amnion, and yolk sac) is
the *allantois*. What is its significance?

⁵

4. Define *decidua*. Name and describe each of its portions (Figure 24-10 in the
text). Which part becomes the "maternal part" of the placenta?

What are *chorionic villi?*

5. Name the structures that comprise the umbilical cord: (*One? Two?*) umbil-

ical artery(-ies), (*one? two?*) umbilical vein(s), and _____

_____ jelly.

6. Define:
a. Afterbirth

b. Umbilicus

D. Developmental anatomy of body systems

1. Complete the table about the development of the integumentary system.

Structure	Which Layer Forms	Type of Tissue Which Forms Structure	Time Period (Month)
a. Epidermis			Second to fourth
b. Nails		Primary nail field	
c. Hair follicles		Stratum basale	
d. Oil glands			
e. Sweat glands			Fourth
f. Connective tissue and blood vessels of skin	Mesoderm		
g. Dermis		Mesenchyme	

6 2. Complete this exercise about the skeletal system.

a. All bones develop from _____ cells derived from (*ectoderm? mesoderm? endoderm?*).

b. Most bones form from _____ tissue which develops from mesenchyme. A few bones, namely the flat skull bones, develop by the process of (*endochondral? intramembranous?*) ossification.

c. Limb buds appear and ossification of the extremities begins during the (*second? fourth? sixth?*) month of development.

d. Remnants of the notochord persist during life as parts of the (*bodies of vertebrae? cartilaginous discs between vertebrae?*).

7 3. Complete the following exercise about muscle development.

a. As the mesoderm develops, it forms dense columns on either side of the developing nervous system. These become segmented into blocks called

_____. The first pair of somites appears on the (*fifth? tenth? twentieth? hundredth?*) day of development. Within ten days, a total of (*11? 22? 44?*) somites are present.

b. (*All? Most?*) of the skeletal muscles of the body develop from the somites. Muscles that do not develop from somites include those in the

_____ and in the _____.

c. Each somite differentiates into three regions. Name these and tell what tissue develops from each.

d. Like skeletal muscle, visceral and cardiac muscle develop from (*ectoderm? mesoderm? endoderm?*).

4. Complete this exercise about the development of the cardiovascular system. $\boxed{8}$

   a. Unlike some developing animals, human embryos have (*much? little?*) yolk for nourishment. Blood and blood vessels begin to form early in development, within the (*third? sixth? ninth?*) week.

   b. Like bone tissue, blood and vessels form from embryonic tissue called _____, derived from _____-derm. Spaces in mesenchymal blood islands form the _____ of blood vessels, while mesenchyme cells form the _____ _____ lining and tunics of blood vessels.

   c. The hearts also formed from _____-derm, begins to develop before the end of the (*first? third? seventh? fifteenth?*) week.

5. Complete this exercise about lymphatic development. $\boxed{9}$

   a. The lymphatic system begins development by the end of the _____ week. It is formed from the _____-derm primary germ layer.

   b. Lymph vessels are derived from lymph sacs which arise from developing (*arteries? veins?*). When these sacs are invaded by mesenchymal cells, they are converted into lymph _____.

6. Arrange the following structures in correct order according to their appearance in the development of the nervous system. Write letters on lines provided. _____ _____ _____ $\boxed{10}$
   G. Neural groove
   P. Neural plate
   T. Neural tube

7. Complete the table below describing development of parts of the brain.

Primary Vesicles	Secondary Vesicles	Structures Formed
a.	1. Telencephalon 2. Diencephalon	
b. Midbrain (mesencephalon)		
c.	1. 2.	Pons, cerebellum

8. The nervous system and many parts of the sense organs are formed from _____-derm.

11 9. Complete this exercise about ear and endocrine development.

   a. The middle ear and eustachian tube develop from the first _____ pouch.

   b. Over the pharyngeal pouches, branchial grooves form. What ear structures develop from the first branchial groove?

   c. The thyroid gland develops at the level of the (*first? second? third? fourth?*) pair of pharyngeal pouches. Parathyroids form from the _____ and _____ pouches. The thymus gland arises from the _____ pouch.

   d. The pituitary gland is derived from the _____-derm. The adrenal cortex arises from the _____-derm, while the adrenal medulla develops from the _____-derm.

e. One endocrine gland that develops from the endoderm (from part of the foregut) is the _____ .

10. Complete this exercise about development of the respiratory system. $\boxed{12}$

   a. The laryngotracheal bud is derived from the _____-derm of the foregut. The proximal end becomes the _____, the middle portion becomes the _____, while the distal part develops into the _____.

   b. The alveoli of the lungs develop after the (*third? sixth? eighth?*) month.

   c. Smooth muscle, cartilage, and connective tissue of bronchial tubes and pleura form from _____-derm.

11. The epithelial lining of the digestive tract is formed from (*ectoderm? mesoderm? endoderm?*), whereas the smooth muscle and connective tissue of this system arises from _____-derm. $\boxed{13}$

12. Complete the table describing the development of digestive organs.

Part of Gut	Organs Formed
a. Foregut	
b. Midgut	
c.	Part of transverse colon, descending and sigmoid colon, rectum, and anus.

13. Describe the development of the salivary glands, liver, gallbladder, and pancreas.

14. Describe the development of the kidneys and ureters, including the roles of the pronephros, mesonephros, and metanephros.

15. Define cloaca, and name two organs which it forms.

14 16. Complete this exercise about development of the reproductive system.

a. The gonads, like the kidneys, develop from _____.

By the end of the _____ week, the gonads are clearly differentiated into ovaries and testes.

b. The Mullerian ducts are structures that persist in the (*male? female?*) reproductive system. These become the organs named _____, _____, and _____.

17. Contrast development of male and female reproductive organs by completing the following table.

Fetal Structure	Structure Formed in Male	Structure Formed in Female
a. Genital tubercle	Penis	
b.		Labia minora
c. Labioscrotal swelling		

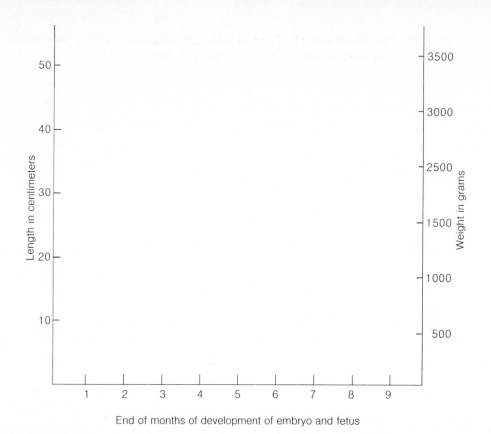

Figure LG 24-1 Graph for demonstrating comparison of growth in length and weight of embryo/fetus. Complete as directed.

18. Refer to Exhibit 24-14. On Figure LG 24-1, graph the length of the embryo or fetus during development. In contrasting color, graph weights of the embryo and fetus. After completing the graph, you can draw the following conclusion: the embryo/fetus grows relatively more in (*length? weight?*) during the early months of development, while in the final months, growth in (*length? weight?*) predominates.

[15]

E. Hormones of pregnancy, parturition and labor, adjustments at birth, lactation   **(pages 699–703)**

1. Complete this exercise about the hormones of pregnancy.   [16]

   a. During pregnancy, the level of progesterone and estrogens must remain (*high? low?*) in order to support the endometrial lining. During the first four months or so, these hormones are produced principally by the

   _____ located in the _____,

   under the influence of the tropic hormone _____.
   (It might be helpful to review the section on endocrine relations in Chapter 19.)

   b. HCG reaches its peak between the _____ and

   _____ month. Since it is present in blood, it will be

   filtered into _____ where it can readily be detected as an indication of pregnancy.

c. The _____ serves as a temporary endocrine organ, producing progesterone and estrogens. It produces these hormones from about the _____ month until _____ .

d. Explain why a pregnant woman does not menstruate while she is pregnant.

e. A woman cannot conceive another child while she is pregnant. The (*high? low?*) level of estrogen and progesterone act to (*stimulate? inhibit?*) FSH. Consequently, she (*will? will not?*) ovulate.

2. Do the following hormones stimulate (S) or inhibit (I) uterine contractions during the birth process?

   17

   _____ a. Progesterone

   _____ b. Estrogen

   _____ c. Oxytocin

3. Contrast these pairs of terms.
   a. Labor/parturition

   b. False labor/true labor

4. What is the "show" produced at the time of birth?

5. Briefly describe the three phases of labor.

6. What adjustments must the newborn make at birth as it attempts to cope with its new environment? Describe adjustments of: respiratory system, heart and blood vessels (review the section on fetal circulation in Chapter 12), and blood.

7. Complete this exercise about lactation. $\boxed{18}$

   a. The major hormone promoting lactation is _____,

   which is secreted by the _____. During pregnancy the level of this hormone increases somewhat, but is inhibited by the

   three hormones _____, _____,

   and _____.

   b. Following birth and the loss of the placenta, a major source of two of

   these hormones (_____ and

   _____) is gone, so prolactin (*increases? decreases?*) dramatically.

   c. The sucking action of the newborn facilitates lactation in two ways. What are they?

   d. What hormone stimulates milk letdown? _____

8. What is *colostrum?*

(pages 703–708) **F. Amniocentesis, birth control, key medical terms**

    1. Describe each of the following birth control methods. Discuss the effectiveness of each.

      a. Vasectomy

      b. Tubal ligation, including laparoscopy

      c. Abstinence during time ovulation is predicted

      d. Condom

      e. Diaphragm

      f. IUD

      g. Spermicides

      h. Oral contraceptives

    2. Define *amniocentesis*. Explain the clinical significance of this procedure in diagnosing chromosomal disorders such as *Down's syndrome*.

3. Define the following terms.
    a. Abortion

    b. Cesarean section

    c. Hermaphroditism

    d. Karyotype

Answers to Numbered Questions in Learning Activities

1.   (a) Months, seminiferous tubules, FSH (and FSHRF indirectly), toward. (b) Spermatogonia, 46, diploid. (c) Chromatids, primary spermatocytes. (d) Homologous, father's sperm, mother's ovum, 23, tetrad, crossing over. (e) Mitotic, crossing over. (f) Secondary spermatocytes, spermatids, spermatozoa, do not.

2.   (a) Identical, 46, 2. (b) Different, 23, 4 spermatozoa. (c) Different, 23, 4 (1 ovum and 3 polar bodies).

3.   (a) Parasympathetic, sacral cord, erect. (b) Sympathetic, emission, ejaculation.

4.   (a) Amniotic. (b) Disc, ecto, amniotic, skin, amnion, bag of waters. (c) Endoderm, gut, yolk, is not. (d) Mesoderm, chorion, blastocoel, coelom, chorion.

5.   Forms umbilical blood vessels.

6.   (a) Mesenchymal, mesoderm. (b) Cartilage, intramembranous. (c) Second. (d) Cartilaginous discs between vertebrae.

7.   (a) Somites, twentieth, 44. (b) Most, head and extremities. (c) Myotome—most skeletal muscles; dermatome—connective tissue including dermis; sclerotome—vertebrae. (d) Mesoderm.

8.   (a) Little, third. (b) Mesenchyme, meso-, lumens, endothelial. (c) Meso-, third.

9   (a) Fifth, meso-. (b) Veins, nodes.

10   P G T.

11   (a) Pharyngeal. (b) External ear and external auditory canal. (c) Second; third and fourth; third. (d) Ecto-, meso- ecto-. (e) Pancreas.

12   (a) Endo-, larynx, trachea, bronchi and lungs. (b) Sixth. (c) Meso-.

13   Endoderm, meso-.

14   (a) Intermediate mesoderm, eighth. (b) Female, vagina, uterus, and uterine tubes.

15   Length, weight.

16   (a) High, corpus luteum, ovary, HCG. (b) First, third, urine. (c) Placenta, second, birth. (d) Estrogen and progesterone levels are high; if they should drop, "spotting" (slight bleeding) could occur, signaling a possible miscarriage. (e) High, inhibit, will not.

17   (a) I (it is *pro*-gestation, anti-birth). (b) S. (c) S.

18   (a) Prolactin, anterior pituitary, estrogen, progesterone, PIF. (b) Estrogen, progesterone, increases. (c) Maintains prolactin level by inhibiting PIF, and stimulates release of oxytocin. (d) Oxytocin.

MASTERY TEST: Chapter 24

Questions 1–3: Arrange the answers in correct sequence.

____ ____ ____ ____   1. From most superficial to deepest (closest to embryo):
    A. Amnion
    B. Amniotic cavity
    C. Chorion
    D. Decidua

____ ____ ____ ____ ____   2. In order of development:
    A. Primary oocyte
    B. Secondary oocyte
    C. Oogonium
    D. Ovum
    E. Ootid

____ ____ ____ ____ ____   3. Stages in development:
    A. Morula
    B. Blastocyst
    C. Zygote
    D. Fetus
    E. Embryo

Questions 4–11: Choose the one best answer to each question.

____   4. All of the following structures are developed from mesoderm EXCEPT:
    A. Aorta   B. Biceps muscle   C. Heart
    D. Humerus   E. Sciatic nerve   F. Neutrophil

_____ 5. Which part of the structures surrounding the fetus is developed from maternal cells (not from fetal)?
   A. Allantois   B. Yolk sac   C. Chorion   D. Decidua
   E. Amnion

_____ 6. All of the following events occur during the first month of embryonic development EXCEPT:
   A. Endoderm, mesoderm, and ectoderm are formed.
   B. Amnion and chorion are formed.
   C. Implantation occurs.
   D. Ossification begins.
   E. The heart begins to beat.

_____ 7. Implantation of a developing individual (blastocyst stage) usually occurs about _____ after fertilization.
   A. Three weeks   B. One week   C. One day
   D. Seven hours   E. Seven minutes

_____ 8. Choose the one FALSE statement.
   A. Excretion of estrogen in the urine serves as the basis for most pregnancy tests.
   B. Most cells of the body are diploid.
   C. The erect state of the penis is caused by parasympathetic nerve impulses.
   D. The umbilical cord contains two umbilical arteries and one umbilical vein.

_____ 9. All of the following structures are formed from mesoderm EXCEPT:
   A. Lymph vessels   B. Blood   C. Cartilage
   D. Muscles   E. Cranial and spinal nerves

_____ 10. Choose the one TRUE statement.
   A. Maternal blood mixes with fetal blood in the placenta.
   B. The stomodeum develops into the oral cavity, whereas the proctodeum forms the anus.
   C. Most skeletal muscles form from the sclerotome portions of somites.
   D. Both the pronephros and mesonephros persist throughout life as parts of the kidneys.

_____ 11. Which of the following processes occurs in both meiosis and mitosis?
   A. Two nuclear divisions   B. Crossing over
   C. Replication of DNA   D. Pairing of homologous chromosomes
   E. Reduction of chromosomes from diploid to haploid number

Questions 12–15: Circle T (true) or F (false). If the statement is false, change the underlined word or phrase so that the statement is correct.

T   F   12. Of the primary germ layers, <u>only the ectoderm forms part of a fetal membrane.</u>

T   F   13. <u>Spermatogonia and oogonia both</u> continue to divide throughout a person's lifetime to produce new <u>primary spermatocytes or primary oocytes.</u>

T   F   14. Oral contraceptives are considered <u>more</u> effective in preventing pregnancy than diaphragms or condoms.

T   F   15. In the normal birth process the delivery of the placenta occurs <u>after</u> the delivery of the baby.

25

Surface Anatomy

Throughout the preceding chapters, you have studied the anatomy of all the body systems. Much information about the body can be gained by viewing superficial structures of the body or by feeling (palpating) these surface areas. In this chapter you will define and consider the significance of surface anatomy (Objective 1) and then identify major aspects of surface anatomy of the entire body: head, neck, trunk, and extremities (2–7).

Topics Summary

A. Head, neck, and trunk
B. Extremities

Objectives

1. Define surface anatomy.
2. Identify the principal regions of the human body.
3. Describe the surface anatomy features of the head.
4. Describe the surface anatomy features of the neck.
5. Describe the surface anatomy features of the trunk.
6. Describe the surface anatomy features of the upper extremities.
7. Describe the surface anatomy features of the lower extremities.

(pages 711–716)
A. Head, neck, and trunk

1. 1. Complete the table relating common terms to anatomical terms.

Common Term	Anatomical Term
a.	Cephalic region
b. Brain case or skull	
c.	Occiput
d. Eyebrows	
e.	Palpebrae
f. Anterior portion of mandible	
g. Cheek	
h.	Auricular region

2. 2. Complete this exercise about the head.

a. The head is divided into two regions; these are _____

_____ and _____.

b. Name the four aspects of the cranium that can be observed from the surface.

3. Using a mirror, identify the following areas of your face: *infraorbital, parotid-masseteric,* and *zygomatic.*

4. Refer to Exhibit 25-1 in your text. Using those figures and/or a partner, locate as many as you can of the numbered structures of the eye, ear, nose, and lips.

3. 5. Complete this exercise about the neck. Refer to Exhibit 25-2.

a. Name the four regions of the neck.

b. Which of these regions contains the thyroid and cricoid cartilages and the hyoid bone?

c. Each lateral cervical region can be divided into two triangles by the diagonally positioned _____ muscle. The anterior triangle thus formed is bordered superiorly by the _____ _____ and medially by the _____ _____. The posterior (lateral) triangle is bordered inferiorly by the _____ and posteriorly by the anterior border of the _____ muscle.

6. Identify the following landmarks on a partner or wall chart.
   a. On the dorsum: *seventh cervical vertebral spine; spine, acromion,* and *borders of the scapula; latissimus dorsi, trapezius,* and *teres major muscles.*
   b. On the chest: *clavicles, sternum, jugular notch, sternal angle; pectoralis major* and *serratus anterior muscles.*
   c. On the abdomen: *umbilicus, rectus abdominis, tendinous intersections,* and *linea alba.*

B. Extremities

(pages 716–723)

1. Complete the table relating common terms to anatomical terms.

4

Common Term	Anatomical Term
a. Armpit	
b.	Brachium
c. Elbow	
d. Forearm	
e.	Manus
f. Wrist	
g.	Metacarpus
h. Fingers and toes	
i. Buttocks	
j.	Femoral region
k. Knee	
l.	Lateral malleolus

5 2. Circle the answer which best fits each descripton in this exercise.
   a. Frequent site of intramuscular injections in the upper extremity: (*flexor carpi ulnaris? trapezius? deltoid?*).
   b. Muscle producing bulge on the superior and lateral aspect of the forearm: (*brachioradialis? coracobrachialis? brachialis?*).
   c. Slight elevation at the lateral end of the clavicle: (*sternal angle? manubrium? acromioclavicular joint?*).
   d. Groove posterior to the medial epicondyle: (*site of the inner ankle? site of the ulnar nerve? site of the radial nerve?*).
   e. Location of a vein often used for intravenous therapy; on anterior surface of elbow joint: (*cubital fossa? lateral epicondyle? popliteal fossa?*).
   f. "Anatomical snuffbox": (*depression between tendons of two muscles that move the thumb? depression between tendons of two muscles that move the forearm?*).
   g. "Knuckles": (*proximal ends of proximal phalanges of the hand? distal ends of second through fifth metacarpals?*).
   h. Protuberance at distal end of the ulna: (*coronoid process? olecranon process? styloid process?*).
   i. Superior border of the buttock: (*iliac crest? greater trochanter? ischial tuberosity?*).
   j. Frequently used as an injection site of insulin in diabetics: (*vastus intermedium? vastus lateralis? adductor longus?*).
   k. Bony prominence below the patella: (*patellar ligament? Achilles tendon? tibial tuberosity?*).
3. On a partner or on yourself, locate as many as you can of the anatomical structures listed in answers to the above exercise.
4. Indicate if the following structures can be palpated on the anterior (A) or
6 the posterior (P) surface of the body:

_____ a. Gastrocnemius muscle

_____ b. Popliteal fossa

_____ c. Biceps brachii

_____ d. Achilles tendon

_____ e. Rectus femoris

_____ f. Tendon of palmaris longus muscle

Answers to Numbered Questions in Learning Activities

1 (a) Head. (b) Cranium. (c) Occipital region or base of skull. (d) Supercilia. (e) Eyelids. (f) Mental region. (g) Buccal region. (h) Ear.

2 (a) Cranium, face. (b) Frontal, parietal, temporal, occipital.

3 (a) Anterior cervical, two lateral cervicals, and posterior region (nucha). (b) Anterior cervical. (c) Sternocleidomastoid, mandible, midline, clavicle, trapezius.

4 (a) Axilla. (b) Arm. (c) Cubitus. (d) Antebrachium. (e) Hand. (f) Carpus. (g) Palm. (h) Digits or phalanges. (i) Gluteal region. (j) Thigh. (k) Genu. (l) Outer ankle.

5   (a) Deltoid. (b) Brachioradialis. (c) Acromioclavicular joint. (d) Site of the ulnar nerve. (e) Cubital fossa. (f) Depression between tendons of two muscles that move the thumb. (g) Distal ends of second through fifth metacarpals. (h) Styloid process. (i) Iliac crest. (j) Vastus lateralis. (k) Tibial tuberosity.

6   (a) P. (b) P. (c) A. (d) P. (e) A. (f) A.

MASTERY TEST: Chapter 25

Questions 1–8: Arrange the answers in correct sequence.

_____ _____ _____   1. From anterior to posterior:
               A. Occiput
               B. Frontal region
               C. Parietal region

_____ _____ _____   2. From superior to inferior:
               A. Palpebrae
               B. Oral region
               C. Infraorbital region

_____ _____ _____   3. From superior to inferior:
               A. Xiphoid process
               B. Umbilicus
               C. Hyoid bone

_____ _____ _____   4. From proximal to distal:
               A. Antebrachium
               B. Brachium
               C. Carpus

_____ _____ _____   5. From proximal to distal:
               A. Phalanges
               B. Proximal transverse flexure lines of palm
               C. Distal transverse flexure lines of palm

_____ _____ _____   6. From medial to lateral:
               A. Acromioclavicular joint
               B. Deltoid muscle
               C. Sternal angle

_____ _____ _____   7. From anterior to posterior of the ear:
               A. Antihelix
               B. Tragus
               C. Helix

_____ _____ _____   8. From medial to lateral of the eye:
                A. Medial canthus
               B. Iris
               C. Lateral canthus

Questions 9–15: Circle T (*true*) or F (*false*). If the statement is false, change the underlined word or phrase so that the statement is correct.

T   F   9. The ulnar nerve lies in a groove behind the <u>medial epicondyle</u>.

T   F   10. The <u>ischial spine</u> bears most of the weight of a person in sitting position.

T   F   11. All of the following surface features are located on the lower extremities: <u>tibial tuberosity, calcaneal tendon, lateral malleolus</u>.

T   F   12. The triceps brachii is located on the <u>posterior</u> surface of the body.

T   F   13. The epicondyles are markings on the <u>ulna</u>.

T   F   14. The trapezius muscle forms the <u>posterior</u> border of the <u>posterior triangle</u> of the <u>lateral cervical region</u>.

T   F   15. The linea alba is located on the <u>posterior</u> surface of the <u>thorax</u>.

Answers to Mastery Tests

CHAPTER 1

Multiple Choice

1. A
2. C
3. D
4. B
5. B
6. D
7. E
8. B

True-False

9. T
10. F. Stand erect facing observer, arms at sides, palms forward
11. T
12. F. Cell (or cellular)
13. F. Two-dimensional
14. T
15. T

CHAPTER 2

Multiple Choice

1. B
2. E
3. A
4. E
5. E
6. D
7. A
8. A

True-False

9. F. Phospholipid molecules
10. F. Some of the characteristics of
11. F. Flagella
12. T
13. T
14. F. Continually
15. F. Nuclear division (Cytokinesis is cytoplasmic division.)

CHAPTER 3

Multiple Choice

1. C
2. D
3. B
4. A

Matching

5. Cartilage
6. Dense connective tissue
7. Stratified squamous epithelium
8. Loose connective tissue
9. Simple squamous epithelium

True-False

10. T.
11. F. Poor conductor of heat and therefore reduces heat loss (provides insulation)
12. T
13. T
14. F. Stratified
15. T

CHAPTER 4

Multiple Choice

1. A
2. B
3. B
4. E
5. D

Arrange

6. C B A
7. A C B
8. C B A
9. B A C

True-False

10. F. Composed of different kinds of cells
11. F. Melanin
12. T
13. T
14. F. Two to five years
15. F. Only the dermis

CHAPTER 5

True-False

1. F. Children
2. F. Mineral salts, and one third is due to collagenous fibers (The very few cells contribute little to bone weight.)
3. T
4. F. End of the bone (often bulbous)
5. F. Protrudes through the skin
6. T

7. F. Fluid from blood vessels in Haversian canals, but not blood itself
8. F. Only in origin
9. T
10. T
11. F. Hyaline cartilage; a few start out as fibrous membranes (Originally all bones begin as mesenchyme.)
12. T

Arrange

13. B C A
14. A C B
15. A C B

CHAPTER 6

Multiple Choice

1. C
2. B
3. B
4. B
5. D
6. A
7. C
8. C
9. C

Arrange

10. A B C
11. B A C
12. C A B

True-False

13. F. Intercostal space
14. T
15. T

CHAPTER 7

Multiple Choice

1. E
2. A
3. D
4. C

Arrange

5. A C B
6. C B A
7. B C A

True-False

8. F. Do not
9. T
10. F. Shallower and more oval
11. T
12. T

13. F. Only tibia and talus
14. F. No other bone (It is used for muscular attachment only.)
15. T

CHAPTER 8

True-False

1. T
2. T
3. F. Do not cover surfaces of articular cartilages
4. T
5. T
6. F. Flexion
7. F. Much less; is only about 0.5 ml in a large joint
8. T
9. T
10. F. Less

Arrange

11. B A C
12. C B A

Multiple Choice

13. A
14. C
15. B

CHAPTER 9

True-False

1. T
2. F. Neither A bands nor I bands
3. F. Dense connective tissue (which may surround skeletal muscle)
4. F. Slow
5. F. Deficient acetylcholine or excess cholinesterase production
6. T
7. F. Atrophy
8. T

Arrange

9. A C B
10. C A B
11. C A B
12. A B C

Multiple Choice

13. C
14. E
15. D

CHAPTER 10

Arrange

1. B C A
2. A B C

3. T
4. F. Antagonists
5. F. Shape (trapezoid)
6. T
7. F. Posterior, except for extensors of leg and foot
8. F. Two heads of origin; but origins of biceps brachii are on the scapula, and origins of biceps femoris are on ischium and femur

Multiple Choice

9. D
10. C
11. D
12. C
13. B
14. D
15. A

CHAPTER 11

Multiple Choice

1. A
2. C
3. D
4. B
5. C
6. A
7. C
8. D
9. C
10. B

True-False

11. T
12. T
13. T
14. F. Shorter
15. F. 47 ml (males) to 42 ml (females)

CHAPTER 12

Arrange

1. C A B D
2. A B C D
3. A B C D E
4. C E D A B

Multiple Choice

5. D
6. C
7. B
8. B
9. D
10. D
11. B
12. C

13. B
14. C
15. A

CHAPTER 13

Arrange

1. A C B D
2. D C B A E

Multiple Choice

3. E
4. C
5. D
6. B
7. D
8. C

True-False

9. T
10. T
11. T
12. F. Less
13. F. Puberty
14. F. In groups
15. T

CHAPTER 14

Arrange

1. C A B
2. A C B

Multiple Choice

3. A
4. C
5. C
6. E

True-False

7. F. May be myelinated since neuroglia myelinate CNS fibers
8. T
9. T.
10. F. Faster
11. F. Spinal nerves (as well as some other structures; but not the brain)
12. T
13. F. Anesthetics depress and caffeine facilitates
14. F. Only neurons
15. T

CHAPTER 15

Arrange

1. B C A
2. C B A D
3. D B C A E

4. E (C and D)
5. D
6. C
7. A
8. B

True-False

9. F. Intact cell body and a neurilemma
10. T
11. T
12. F. Ipsilateral, meaning "same-sided"
13. F. Are not
14. F. Brachial
15. F. Above C5 level

CHAPTER 16

Multiple Choice

1. E
2. B
3. A
4. A
5. E
6. C
7. B

Arrange

8. B A C
9. C A B
10. A C B
11. A B D C

True-False

12. T
13. T
14. T
15. T

CHAPTER 17

Multiple Answers

1. C, D, E
2. A, E, F
3. A, C, E, G
4. A, B, C, D
5. B

Multiple Choice

6. E
7. C
8. D

True-False

9. T
10. T
11. T
12. F. Dependent on
13. T
14. T
15. F. Some (or most)

CHAPTER 18

Arrange

1. A C B
2. C D A B
3. C D A B
4. D A C B
5. D B A C
6. B C D E A
7. A E C B D
8. A D E B C
9. B C A D E
10. A B E C D

Multiple Choice

11. E
12. D
13. A
14. A
15. B

CHAPTER 19

Multiple Choice

1. C
2. D
3. E
4. C
5. E
6. E

True-False

7. T
8. F. Hypothalamus and affect the anterior pituitary
9. T
10. F. Sympathomimetic
11. T
12. F. Hypophysis (pituitary)
13. F. Adenohypophysis
14. F. Prolactin and oxytocin
15. F. Diabetes insipidus

CHAPTER 20

Arrange

1. A C B
2. D C E B A
3. A E C B D
4. A E C B D
5. D C E B A F

6. B
7. E
8. A
9. B

True-False

10. T
11. F. Cuboidal Type II cells of alveolar walls
12. F. The bifurcation
13. F. Right, left, right, left
14. T
15. F. Exchange of gases between the atmosphere and the lungs

CHAPTER 21

Multiple Choice

1. B
2. C
3. B
4. A
5. A
6. E

Arrange

7. A D B C
8. B D C A
9. E D B A C
10. A D C B E
11. A B D C E

True-False

12. T
13. F. No digestive enzymes, but it does produce mucus
14. F. Villi, microvilli, and goblet cells (Rugae are in wall of the stomach.)
15. T

CHAPTER 22

Arrange

1. A B C
2. A C B D
3. E B A D C
4. C E A B D
5. A C E B D

Multiple Choice

6. B
7. D
8. B
9. A
10. C
11. D
12. B

True-False

13. F. A kidney to the bladder
14. T
15. F. Incontinence

CHAPTER 23

Arrange

1. B C A
2. A B C
3. B A C
4. C D B A
5. B E D A C
6. B C D A E
7. F C B D A E

Multiple Choice

8. B
9. E
10. C

True-False

11. F. Cooler
12. T
13. F. Amount of glandular tissue
14. T
15. F. Testicular arteries which are branches of the aorta

CHAPTER 24

Arrange

1. D C A B
2. C A B E D
3. C A B E D

Multiple Choice

4. E
5. D
6. D
7. B
8. A
9. E
10. B
11. C

True-False

12. F. All three layers form part of the fetal membrane (Mesoderm forms part of chorion plus allantois; endoderm forms yolk sac.)
13. F. Spermatogonia; primary spermatocytes (Oogonia do not divide after birth of female baby.)
14. T
15. T

CHAPTER 25

Arrange

1. B C A
2. A C B
3. C A B
4. B A C
5. B C A
6. C A B
7. B A C
8. A B C

True-False

9. T
10. F. Ischial tuberosity
11. T
12. T
13. F. Humerus
14. T
15. F. Anterior, abdomen